The Laboratory Environment

The Laboratory Environment

Edited by

Rupert Purchase

Environment Group, The Royal Society of Chemistry

ROYAL
SOCIETY OF
CHEMISTRY

Based on the proceedings of a one-day symposium organized by the Royal Society of Chemistry Environment and Toxicology Subject Groups in association with the Health and Safety Group of the Society of Chemical Industry on The Laboratory Environment: Working with Dangerous Substances held on 30 March 1993 at Burlington House, London.

Special Publication No. 136

ISBN 0-85186-605-0

A catalogue record for this book is available from the British Library

Published by The Royal Society of Chemistry,
Thomas Graham House, Science Park, Cambridge
CB4 4WF

Printed in Great Britain by Redwood Books, Trowbridge, Wiltshire

Preface

At the 1890 *Benzolfeier* to honour his work on the structure of benzene, August Kekulé is reported to have remarked:

'If you want to become a chemist, so Liebig told me, when I worked in his laboratory, you have to ruin your health. Who does not ruin his health by his studies, nowadays will not get anywhere in Chemistry'.[1]

Mercifully in the 1990s these opinions are extinct, and safe techniques and advice for handling chemicals in a laboratory are firmly in place — partly as a result of legislation[2,3] and improvements in ventilation design,[4] and partly through the efforts of professional organizations for chemists such as The Royal Society of Chemistry.[5]

Scientists, occupational health physicians and hygienists, and engineers, among others, have contributed to our knowledge of how to work safely with potentially hazardous chemicals and biological materials in chemical laboratories, dispensaries, and similar workplaces. This book describes how practitioners from these disciplines can interact to provide a safe environment for those, mainly chemists, working within such laboratories and for those who may be affected as a consequence of a laboratory's activities. The main areas covered are **Surveillance** (Chapters 1-3); **Handling and Disposal** (Chapters 4-11); and **Control of Exposure** (Chapters 12-13).

Because the range of substances handled by those with a training in chemistry and related subjects is much wider than encountered 20 or 30 years ago (let alone in the 19th century), the opportunity has been taken to include accounts of handling radioactive materials (Chapter 8) and infectious waste (Chapter 9) in the remit for this book.[6]

Among the themes discussed are: appropriate health surveillance measures for laboratory and other staff (Chapters 2 and 7); the classification, handling, and disposal of carcinogens (Chapters 5-7); the recycling of laboratory waste (Chapter 10); literature sources for assessing reactive chemical hazards (Chapter 11); and the relationship between client and contractor in the design and construction of laboratory areas (Chapters 12 and 13).

Underpinning any laboratory safety policy is the will to act safely. This includes the adoption of good safety management practices as well as a clear understanding of safety legislation. Advice on complying with current and forthcoming UK and EC health and safety regulations and guidelines forms a central part of this book (Chapters 3, 4, 8, and 9).

The summary of a 1983 RSC symposium on health and safety in the chemical laboratory[7] included a request for chemical societies to:

- Make the general public more aware that professional chemists are actively engaged in making chemistry safer; and

- Counter and discourage adverse media comments regarding health and safety.

Other recommendations from that meeting called for greater collaboration on health and safety among professional bodies in the fields of chemistry and toxicology, and identified a need for:

- Reduced premiums to be charged by insurance companies to organizations with good health and safety records;

- Common procedures for the identification of, and reporting on, the incidence of ill-health from work-related disease; and

- An internationally accepted definition of a carcinogen.

An additional aim of this book has been to try to address and comment on these issues with the benefit of the intervening decennium.

This book is based on the proceedings of a symposium held in London in March 1993, and organized jointly by the Environment and Toxicology Groups of The Royal Society of Chemistry in association with the Health and Safety Group of the Society of Chemical Industry. I thank Mr. Mervyn Richardson of the RSC Toxicology Group and Mrs. Pauline Sim of Gascoigne Secretarial Services for their help with the organization of this meeting, and the members of the RSC Environment Group Committee for their support. I am indebted to Pauline Sim for retyping the contributors' manuscripts, and her assistance in the production of this book.

Footnotes and References

1. A. Kekulé, *Ber.*, 1890, **23**, p. 1308; quoted in a review of the life and work of Heinrich Wieland (who had an apparently equally phlegmatic attitude to the health of his students who isolated toxic fungal metabolites), B. Witkop, *Angew. Chem. Int. Ed.*, 1977, **16**, 559.

2. (a) N. Selwyn, 'Law of Health and Safety at Work', Butterworths, London, 1982; (b) 'Tolley's Health and Safety at Work Handbook 1992-1993', ed. M. Dewis, Tolley Publishing Company, Croydon, 1992.

3. 'Safe Practices in Chemical Laboratories', Royal Society of Chemistry, London, 1989, [Section F of this booklet summarises the main provisions of the Health and Safety at Work Act 1974, subsequent Regulations made under the Act

(including the Control of Substances Hazardous to Health Regulations 1988), and other legislation affecting laboratories].

4. *cf.* (a) A.W. Hofmann, 'The Chemical Laboratories in Course of Erection in the Universities of Bonn and Berlin', W. Clowes and Sons, London, 1866, pp. 38-43 (a description of the 'evaporation niches'); (b) H.E. Roscoe, 'Description of the Chemical Laboratories at the Owens College, Manchester', J.E. Cornish, Manchester, 1881, pp. 3-4; (c) K. Everett and D. Hughes, 'A Guide to Laboratory Design', Butterworths, London, 1975, Chapter 6; (d) L.J. DiBerardinis *et al.*, 'Guidelines for Laboratory Design: Health and Safety Considerations', Second Edition, J. Wiley, New York, 1993.

5. *e.g.* (a) 'Laboratory Handbook of Toxic Agents', ed. C.H. Gray, Royal Institute of Chemistry, London, 1960; Second Edition 1966. Revised Edition, with the title 'Hazards in the Chemical Laboratory', ed. G.D. Muir, Royal Institute of Chemistry, London, 1971; Subsequent Editions 1977, 1981, 1986. Fifth Edition, ed. S.G. Luxon, Royal Society of Chemistry, Cambridge, 1992; (b) 'Safety in the Chemical Laboratory', ed. N.V. Steere, Volumes 1-3, Division of Chemical Education of the American Chemical Society, Pennsylvania, 1967, 1971, 1974; (c) International Union of Pure and Applied Chemistry and International Programme on Chemical Safety, 'Chemical Safety Matters', Cambridge University Press, Cambridge, 1992.

6. Injuries which may result from exposure to chemical and biological hazards and ionizing radiation have been causally linked to a 'transfer of energy in excess of an individual's damage threshold'. Other types of energy or hazard present in a laboratory environment which can cause injury, directly or indirectly, are: mechanical energy; electrical energy; nonionizing radiation; atmospheric pressure differentials; thermal energy; and fire (N.V. Steere, in 'Laboratory Safety: Theory and Practice', ed. A.A. Fuscaldo, B.J. Erlick, and B. Hindman, Academic Press, New York, 1980, pp. 3-28).

7. 'Health and Safety in the Chemical Laboratory — Where do we go from here?' Royal Society of Chemistry, London, 1984.

Rupert Purchase, *Editor*
Haywards Heath, West Sussex, UK.

August 1993

Contributors

Raymond M. Agius, *Department of Public Health Sciences, University of Edinburgh, Teviot Place, Edinburgh EH8 9AG, UK.*

Colin Bloch, *W.F. Johnson & Partners Ltd., 40 Queen Square, Bristol BS1 4QP, UK.*

Marcel Castegnaro, *International Agency for Research on Cancer, 150 Cours Albert-Thomas, 69372 Lyon, Cedex 08, France.*

Peter G.W. Cobb, *Professional Affairs Board, The Royal Society of Chemistry, Burlington House, London W1V 0BN, UK.*

Christopher H. Collins, *'The Ashes', Hadlow, Kent TN11 0AS, UK.*

Michael Hannant, *The Royal Society of Chemistry, Thomas Graham House, Science Park, Milton Road, Cambridge CB4 4WF, UK.*

H. Paul A. Illing, *Toxicology Group, The Royal Society of Chemistry, Burlington House, London W1V 0BN, UK.*

Belinda Kershaw, *AEA Technology Harwell, Didcot, Oxfordshire OX11 ORA, UK.*

M. Gerard Lee, *Mersey Regional Health Authority, Hamilton House, 24 Pall Mall, Liverpool L3 6AL, UK.*

John Salmon, *Rhône-Poulenc Rorer Ltd., Dagenham Research Centre, Rainham Road South, Dagenham, Essex RM10 7XS, UK.*

Michael P. Satchell, *Merck Ltd., West Quay Road, Poole, Dorset BH15 1HX, UK.*

Peter Watts, *BIBRA Toxicology International, Woodmansterne Road, Carshalton, Surrey SM5 4DS, UK.*

Ian Wrightson, *Eagle Star Insurance Co Ltd., 54 Hagley Road, Edgbaston, Birmingham, B16 8QP, UK.*

Anthony Yardley-Jones, *'Cornerfield', Northchurch Common, Berkhamsted, Hertfordshire, HP4 1LR, UK.*

Contents

CHAPTER 1

The Laboratory Environment and The Royal Society of Chemistry

P.G.W. COBB

1 Introduction

Article 3 of the Royal Charter[1] which was granted to The Royal Society of Chemistry (RSC) in March 1980 following the merger of the Chemical Society and the Royal Institute of Chemistry states that:

'The object for which the Society is constituted is the general advancement of chemical science and its application and for that purpose:

i) to foster and encourage the growth and application of such science by the dissemination of chemical knowledge;

ii) to establish, uphold and advance the standards of qualification, competence and conduct of those who practise chemistry as a profession;

iii) to serve the public interest by acting in an advisory, consultative or representative capacity in matters relating to the science and practice of chemistry; and

iv) to advance the aims and objectives of members of the Society so far as they relate to the advancement of the science or practice of chemistry'.

Through learned and professional services to its 43,000 members, through services to Members of Parliament using Parliamentary Advisers and Link MPs, by taking an active part in the consultative procedures with government departments, and by extensive activities in education and publishing,[1] the RSC can claim to be advancing chemical science both within the spirit and the letter of the Royal Charter.

Four current activities of the RSC may be quoted as examples of its specific involvement in the health and safety aspects of the practice of chemistry and the creation of a safe laboratory environment:

• the work of the RSC's Professional Affairs Board and of the Environment, Health and Safety Committee;

- the new edition of the established handbook on laboratory safety — 'Hazards in the Chemical Laboratory';[2]

- the national surveys of the lifestyles and work-styles of RSC members (the 'RECAP' programme), and of the mortality of professional chemists;

- the monthly publication 'Laboratory Hazards Bulletin', (which is described in detail in Chapter 11).

2 The RSC's Professional Affairs Board and the Environment, Health and Safety Committee

The Environment, Health and Safety Committee reports to the Society's Professional Affairs Board and has a basic membership of seven practitioners from the three fields of interest. Because of its wide remit the Committee also co-opts additional members from other parts of the Society, *e.g.* the Environmental Chemistry Group, and elsewhere, as appropriate, to give a total membership of about 20.

Much of the work of the Environment, Health and Safety Committee is quasi-legal and relates to the development of new legislation and the legal implications for RSC members. For example the Committee responds on behalf of the RSC to consultative documents on health, safety and environmental matters sent to the Society by bodies such as the UK Health and Safety Executive, the Department of the Environment, and the British Standards Institution.

Since January 1992 the Committee has prepared over 30 responses to consultative documents including the Health and Safety Commission's consultation papers *Genetically Modified Organisms (GMOS) Proposed New Regulations* and *Draft Proposals for the Chemicals (Hazard Information and Packaging) Regulations* (the CHIP Regulations), and the Royal Commission on Environmental Pollution's study on transport and the environment.

Under the auspices of the Environment, Health and Safety Committee, the RSC has undertaken a series of research contracts on behalf of the Commission of the European Communities (CEC). This collaboration has led to the publication by the RSC of the following books:

'Organo-chlorine Solvents: Health Risks to Workers',[3*]

'Solvents in Common Use: Health Risks to Workers',[4†]

---

*Dichloromethane, chloroform, carbon tetrachloride, 1,2-dichloroethane, 1,1,1-trichloroethane, 1,1,2-trichloroethane, trichloroethylene, perchloroethylene, 1,2-dichloropropane, *p*-dichlorobenzene.
†Acetone, carbon disulphide, diethyl ether, 1,4-dioxane, ethyl acetate, methanol, nitrobenzene, pyridine, toluene, xylene.

'Measurement Techniques for Carcinogenic Agents in Workplace Air',[5]

'Long-term Neurotoxic Effects of Paint Solvents'.[6]

The activities of this Committee are manifest in at least two other areas: the publication of guidance and advisory booklets on laboratory safety which give the Society's considered advice, and the preparation of Professional Briefs which are available free of charge to members.

Three booklets on laboratory safety have been recently issued:

'Safe Practices in Chemical Laboratories'[7] (1989),

'COSHH in Laboratories'[8] (1989),

'Guidance on Laboratory Fume Cupboards'[9] (1990).

Professional Briefs are intended to provide succinct summaries of topical issues in health, safety and the environment. An RSC opinion is presented in these documents, where appropriate. Eleven Professional Briefs have been issued to date,[10] and of these, five are directly related to the laboratory environment:

COSHH and School Science; COSHH in Laboratories; Guidance for Members Involved with 'Healthy Human Volunteer Experiments'; Reproductive Risks of Chemicals at Work; and Classification of Wastes.

3 'Hazards in the Chemical Laboratory'[2]

The RSC publication 'Hazards in the Chemical Laboratory', now in its fifth edition, has developed from an original monograph 'Laboratory Handbook of Toxic Agents', first published by the Royal Institute of Chemistry in 1960.

The new edition of 'Hazards in the Chemical Laboratory', is divided into two parts:

- Eleven introductory chapters written by various authors on laboratory safety covering for example, legal aspects, safety management, fire and radiation protection, electrical hazards, basic toxicology, and first aid procedures.

- The 'yellow pages' section in which the hazardous properties of about 1400 laboratory chemicals are itemized. This section also includes those chemicals listed in the first five volumes of the RSC publication 'Chemical Safety Data Sheets'.[11-15]

Hazardous properties of chemicals are described in the second section under 9 headings for each entry: the commonly accepted name; physical description; risk and safety phrases (taken from the Classification, Packaging, and Labelling Regulations 1984); occupational exposure limits; toxic effects (including any carcinogenicity data relevant to man); hazardous reactions; first aid procedures; fire hazard; and spillage and disposal procedures. The entry for ethylene glycol monobutyl ether (Figure 1) is illustrative.

597. Ethylene glycol monobutyl ether

Colourless liquid; m.p. -74.8 °C; b.p. 168.4 °C; miscible with water.

RISKS
Harmful by inhalation, in contact with skin, and if swallowed — Irritating to respiratory system (R20/21/22,R37)

SAFETY PRECAUTIONS
Avoid contact with skin and eyes (S24/25)

Limit values   MEL long-term 25 p.p.m. (120 mg m^{-3}).

Toxic effects   Potent eye irritant causing intense pain. Inhalation, ingestion, or skin absorption cause changes to the blood system, liver, kidneys, and spleen.

Hazardous   Liable to form explosive peroxides on exposure to air or light
reactions   which should be decomposed before the ether is distilled
to small volume.

First aid   Standard treatment for exposure by all routes (see pages 108-122).

Fire hazard   Flash point 69 °C (open cup); Explosive limits 4-13%; Autoignition temperature 214 °C, Extinguish fires with carbon dioxide, dry chemical powder, or alcohol-resistant foam.

Spillage   See general section.
  disposal

RSC *Chemical Safety Data Sheets Vol. 1, No. 46, 1988* gives extended coverage.

Figure 1   *Health and safety data for ethylene glycol monobutyl ether*
           (Reproduced with permission from 'Hazards in the Chemical Laboratory',
           Fifth edition, ed. S.G. Luxon, Royal Society of Chemistry, Cambridge,
           1992)

4 The RSC Mortality and RECAP Surveys

4.1 Mortality of Professional Chemists in England and Wales 1965 — 1989[16]

This recently published epidemiological study[16] determined the causes of mortality of 4,012 professional chemists, qualified to degree level in chemistry, from an entry group of 14,884 male professional RSC members registered on January 1st 1965.

The deceased, who had practised chemistry for all or part of their working lives, had addresses in England or Wales, and died between January 1st 1965 and December 31st 1989.

The mortality data for chemists were compared with mortality data for males aged 15 or over in Social Classes I and II (professional, technical, and managerial classes) taken from the Registrar General's Decennial Supplement on Occupational Mortality, 1970 — 1972. Data from the period 1970 — 1972 were used because a detailed breakdown on the causes of death had been carried out and could be compared with the RSC data.

The results,[16] which were consistent with other studies on chemists, showed an overall low mortality rate for chemists, with fewer deaths than expected from cancers, but an excess mortality from lymphatic and haematopoietic cancers, in particular leukemias. There was also an excess mortality among chemists from cancers of the duodenum, colon, pancreas, kidney, skin and pleura, cancers and diseases of the nervous system, and mental disorders.

Some of the results of the RSC mortality survey may point to past attitudes among some chemists about handling solvents in the laboratory environment.

The Royal Society of Chemistry is continuing to monitor the mortality of professional chemists, and it is hoped that this work will be supplemented by the results from its RECAP programme. Both surveys have been supported by the CEC.

4.2 RSC Effects of Chemicals Assessment Programme (RECAP)

RECAP is a long-term study by the RSC on the work-style and lifestyle of its members.[17] The survey is based on the results of a questionnaire which seeks information on the health, family medical history, tea, coffee and tobacco consumption patterns, and employment of the participants. Members are also asked about their use of pharmaceutical preparations, and whether they associate any of the health effects recorded with exposure to a particular chemical (or chemicals) in the workplace.

In the most ambitious part of the survey, 800 laboratory chemicals and reagents are listed (including mixtures of isomers and congeners), and the RSC members are requested to identify those chemicals they have encountered at work. Information on the degree of exposure to a particular chemical is not being collated at present.

RECAP was initiated in 1979 and the results of 18,000 questionnaires completed and returned by 1982 have been used to build up a database on chemists of their careers and health, and the chemicals they have used at work. An interim analysis of RECAP data was published by the RSC in 1980.[18]

The latest phase of RECAP, launched in June 1993,[19] seeks to update that information for the period 1982 — 1992 but also includes two aspects not covered in the earlier surveys — details of any disturbance to the reproductive system of the participants experienced during their entire working lives, and clarification of their complete smoking history.

Unpublished RECAP data analysed in 1982 showed that the proportion of non-smokers and reformed ex-smokers among RSC chemists (79%) was greater than for their counterparts in Social Classes I and II (69%).[16] These differences in smoking habits probably account for the reduction in deaths of chemists due to respiratory diseases, including cancer of the trachea, bronchus, and lung.[16]

Ascribing effects on health to a particular lifestyle or to contact with identifiable chemicals at work is clearly a complex task. However, RSC members have shown their eagerness to participate in RECAP in sufficient numbers since 1979 to justify the continuance of this project. Further useful correlations from the statistical analysis of RECAP data can be expected.

5 Conclusion

The Royal Society of Chemistry provides authoritative advice and information on the hazardous properties of chemicals and on the significance for chemists of new and proposed health, safety and environmental legislation. It is also actively investigating on behalf of its members chemicals whose health effects may have hitherto been unnoticed in the laboratory environment and the workplace. The sponsorship of symposia on laboratory safety whose proceedings are recorded in this book and in a previous RSC publication,[20] is also part of the Society's role in health and safety.

By tackling these issues the Society will help to counteract public concern about the use and application of chemicals, and perhaps more importantly, for the future of the RSC and in compliance with its Royal Charter, 'foster and encourage' young people to train in chemistry and promote the benefits of this science in improving the lifestyles and productivity of the whole community.

6 References

1. D.H. Whiffen and D.H. Hey, 'The Royal Society of Chemistry: The First 150 Years', Royal Society of Chemistry, London, 1991, pp. 58-61.

2. 'Hazards in the Chemical Laboratory', Fifth edition, ed. S.G. Luxon, Royal Society of Chemistry, Cambridge, 1992.

3. 'Organo-chlorine Solvents: Health Risks to Workers', Royal Society of Chemistry, London, 1986.

4. 'Solvents in Common Use: Health Risks to Workers', Royal Society of Chemistry, London, 1988.

5. 'Measurement Techniques for Carcinogenic Agents in Workplace Air', Royal Society of Chemistry, Cambridge, 1989.

6. 'Long-term Neurotoxic Effects of Paint Solvents', Royal Society of Chemistry, London, 1990.

7. 'Safe Practices in Chemical Laboratories', Royal Society of Chemistry, London, 1989.

8. 'COSHH in Laboratories', Royal Society of Chemistry, London, 1989.

9. 'Guidance on Laboratory Fume Cupboards', Royal Society of Chemistry, London, 1990.

10. RSC Professional Briefs issued up to July 1993: COSHH and School Science; COSHH in Laboratories; Guidance for Members Involved with 'Healthy Human Volunteer Experiments'; Reproductive Risks of Chemicals at Work; Classification of Wastes; Additives in Food for Human Consumption, General Considerations; Additives in Food for Human Consumption, Food Preservatives; Additives in Food for Human Consumption, Food Irradiation; Additives in Food for Human Consumption, Food Colours; Additives in Food for Human Consumption, Caramel Colours; Contaminated Land.

11. 'Chemical Safety Data Sheets, Volume 1 — Solvents', Royal Society of Chemistry, Cambridge, 1989.

12. 'Chemical Safety Data Sheets, Volume 2 — Main Group Metals and their Compounds', Royal Society of Chemistry, Cambridge, 1989.

13. 'Chemical Safety Data Sheets, Volume 3 — Corrosives and Irritants', Royal Society of Chemistry, Cambridge, 1990.

14. 'Chemical Safety Data Sheets, Volume 4a — Toxic Chemicals (A-L), and Volume 4b — Toxic Chemicals (M-Z)', Royal Society of Chemistry, Cambridge, 1991.

15. 'Chemical Safety Data Sheets, Volume 5 — Flammable Chemicals', Royal Society of Chemistry, Cambridge, 1992.

16. W.J. Hunter, B.A. Henman, D.M. Bartlett, and I.P. Le Geyt, *Am. J. Ind. Med.*, 1993, **23**, 615.

17. D.H. Whiffen and D.H. Hey, 'The Royal Society of Chemistry: The First 150 Years', Royal Society of Chemistry, London, 1991, p. 230.

18. 'RSC Effects of Chemicals Assessment Programme (RECAP) First Report', Royal Society of Chemistry, London, June 1980.

19. B.A. Henman and W.J. Hunter, *Chem. Brit.* 1993, **29**, 677.

20. 'Health and Safety in the Chemical Laboratory — Where do we go from here?', Royal Society of Chemistry, London, 1984.

Editor's Note: Peter Cobb was Chairman of The Royal Society of Chemistry's Professional Affairs Board during 1993.

CHAPTER 2

Health Surveillance of Laboratory Staff

R.M. AGIUS AND A. YARDLEY-JONES

1 Introduction

In his book, 'The Effects of Arts, Trades and Professions on Health and Longevity' published in 1832, Charles Turner Thackrah describes Chemists and Druggists as being exposed to 'various odours, and the evolution of gases, many of which are injurious'. He goes on, 'Hence the persons employed in laboratories are frequently sickly in appearance, and subject to serious affections of the lungs ... few old men are found in the employ.' In the same writing, he suggests that 'Care on the part of the men, and ventilation practised as much as possible, would considerably diminish the effect of the baneful agents'. Thankfully, such working conditions are now almost unheard of in laboratory environments, and advances in control measures and good laboratory practice have ensured that the fate of chemists and laboratory personnel is now not as sinister as in the past!

Health surveillance is usually a part of the comprehensive strategy for dealing with dangerous substances in laboratories and similar workplaces. Often it is no more than a safety net of last resort and even then can have many shortcomings as this chapter will illustrate. All practicable means should first be employed to *assess* the health risks that may be associated with exposure to dangerous substances in the laboratory, to *control* those risks, and to *monitor* the work-environment regularly. The role of *health surveillance* should be seen as one part of this panoply of safety procedures for the laboratory and other workplaces.

If the harmful effects of hazardous substances were known beyond doubt, and if exposure was at demonstrably safe levels, then health surveillance would be unnecessary. However, the health hazards of many substances, especially novel chemical entities, are still unknown. In particular, the quantitative relationship between the airborne concentration of exposure of the individual worker or the dose taken up, and the risk of adverse health effects is often unclear.

Moreover, different individuals may have a wide range of susceptibilities to the same exposure and hence health surveillance may help identify those at higher risk. Measures of exposure may be unavailable or inadequate for some agents with high biological potency, and exposure through routes such as the skin can be impossible to measure; thus measures of uptake or of health effect assume a special importance.

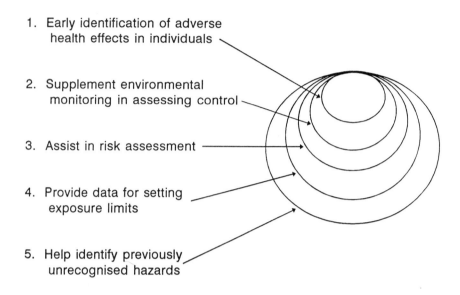

1. Early identification of adverse
   health effects in individuals

2. Supplement environmental
   monitoring in assessing control

3. Assist in risk assessment

4. Provide data for setting
   exposure limits

5. Help identify previously
   unrecognised hazards

Figure 1 *Purposes of health surveillance*

In summary, health surveillance has a number of roles and objectives (linked concentrically as shown in Figure 1) in the creation of a safe working environment:

• Health surveillance should identify, as early as possible adverse health effects resulting from exposure to hazardous substances so as to trigger the appropriate remedial steps, both for the affected individual and others in the workplace.

• Health surveillance can supplement environmental monitoring in assessing the effectiveness of control measures. For example short term exposure to di-isocyanates at concentrations as low as a few parts per billion can provoke asthmatic attacks in some individuals.[1,2] Environmental monitoring on its own is unlikely to be adequate to identify these excursions of exposure.

• Health surveillance may assist in risk assessment by helping to decide whether further steps for the control of exposure, and indeed the continuation of a health surveillance programme, are warranted.

• Health surveillance may provide data for studying the relationship between exposure and adverse effects of health, and hence contribute to the setting of exposure limits. If an organization has a special monopoly of the manufacture or handling of an agent then it should use health surveillance data, in conjunction with environmental monitoring, to try and derive provisional exposure limits.

- Health surveillance may identify previously unrecognized or unproven health hazards. For example, if evidence by analogy with similar substances, or based on animal studies, suggests that a novel chemical entity could present a hazard of sensitization of the human skin or respiratory tract, prospective health surveillance may help determine whether this is so.

2 What is Health Surveillance?

2.1 Definitions

Health surveillance can be understood as a strategy, together with the relevant methods, to detect, as early as feasible, adverse effects on health resulting from exposure to hazardous substances. The term 'Health Surveillance' has been the subject of different, and not always congruent, definitions. Therefore before adopting an operational definition it is useful to present in parallel an outline of the scientific background to the relationship between exposure to dangerous substances and adverse effects on human health (Table 1).

Table 1 *Hazard and health assessment strategies for exposure to hazardous substances*

Chain of events	Means of assessment
Presence of hazardous substances	Hazard assessment
Exposure to hazardous substances	Quantitative measures *e.g.* background or personal monitoring
Absorption of hazardous substances: (and other toxicokinetic effects of biotransformation and excretion)	Biological Monitoring (BM) ['wide definition' of health surveillance]
Toxicodynamic effects of hazardous substances on human biochemistry	Biological Effect Monitoring (BEM) ['wide definition' of health surveillance]
Adverse effect on 'health' *e.g.* through symptoms or other reduction in well being, overt disease, etc.	Health Surveillance ['narrow definition']

The term 'Health Surveillance' in its wide definition is used in legislation, such as the Control of Substances Hazardous to Health (COSHH) Regulations 1988,[3,4] to encompass the techniques of 'Biological Monitoring' (BM) and 'Biological Effect Monitoring' (BEM) (Table 1). For purists 'Health Surveillance' can also be interpreted in a more narrow definition to mean just the measures that identify an adverse effect on health. Measures of uptake (absorption, distribution) or changes in body function not associated with symptoms of ill health or likelihood of disease are excluded from this narrower definition of health surveillance.

2.2 Biological Monitoring (BM) and Biological Effect Monitoring (BEM)

BM and BEM consist respectively of an assessment of the *uptake* or *effect* of chemicals, which are present in the working environment, through appropriate measurements in biological specimens collected from individuals. The measurement can be either that of the substance itself or its metabolite(s) (biological monitoring) or a characteristic reversible biochemical change induced by the chemical (biological effect monitoring).

Such measurements can be made in exhaled air, urine, blood or other biological specimens. However, it should be emphasized that the mere detection of an organic solvent in blood or breath, *e.g.* ethanol or chloroform, or of say, mercury in the urine, does not necessarily mean that health has been harmed. The data have to be interpreted in the light of many variables. Intra-individual and inter-individual differences in tissue levels occurring at the same exposure conditions must be considered.

The advantages and disadvantages of BM and BEM are summarized here. The human response to the same source of exposure varies considerably. This variability has two origins. First, there is the variability associated with differences in the penetration of the chemical to the target organ (toxicokinetics). Secondly, there is the variability associated with differences among individuals in the response and timing of the response of the target organ itself (toxicodynamics). Such variables are responsible for a weakening of the correlation between the level of exposure and the levels of measurement in biological fluids. The reader is referred to texts such as Casarett and Doull's 'Toxicology'[5] for a more detailed account of toxicokinetics and toxicodynamics.

2.2.1 Exposure Level

Tasks and hence potential exposures in the workplace are likely to vary considerably from day-to-day. Target organ concentrations of chemicals with a short biological half-life closely follow the environmental concentrations and, therefore, have a larger variability. On the other hand, levels of chemicals with a long biological half-life show little fluctuation in the target organ.

2.2.2 Route of Exposure

Whilst it is often assumed that chemicals in a laboratory environment are presented to the body *via* the lungs, other routes of intake can be biologically significant. Liquids, and to a much lesser extent gases, can be significantly absorbed through the skin.[6] As a consequence, because air monitoring does not take into account the entire uptake, dermal absorption modifies the relationship between levels of the measurement in biological fluid and the airborne concentration of the chemical in the work environment. The target organ concentration may itself be quite different from the measurement in body fluid which is carried out for health surveillance purposes.

2.2.3 Variability in the Individual

Physical workload varies according to the tasks and the type of industry. Physical workload during exposure can have various effects. Thus it generally increases the rate of uptake of the chemical by increasing pulmonary ventilation and cardiac output, and it may modify the distribution of the chemical in the body tissues by altering the distribution of cardiac output. On the other hand, exercise might help to eliminate volatile chemicals once exposure has ceased. Thus, physical workload may have a complex influence on the toxicokinetics of chemicals.

Individual variability is especially important in the context of BEM. A classic example of biological effect monitoring arises with those organophosphorus compounds and similar substances which act as anticholinesterases, *i.e.* they inhibit the enzyme which destroys the neurotransmitter acetyl choline. The substances themselves can be very difficult or impossible to detect in the human body after exposures small enough not to result in symptoms. Yet it can still be possible to demonstrate a biological effect, namely a reduction in the activity of the cholinesterase circulating in the blood, and this can be presumed to be a surrogate of the biological effect on the enzyme in the nerve endings where it matters more. Of course, there is usually no clear fixed line of demarcation above which a 'biological effect' becomes an 'adverse health effect'.

2.2.4 Confounding Factors

Endogenous factors such as age, sex and disease can affect the metabolism of chemicals. For example, liver disease may be associated with a decreased biotransformation of chemicals, whereas renal disease may impair their elimination *via* the urine.

In a laboratory environment, as in many other chemical processes, one has also to consider that it is unlikely that exposure to a single agent is the norm. Exposures to mixtures are usually underestimated. Such multiple exposures may lead to an interaction between the chemicals and this can cause difficulties in the interpretation of data, *e.g.* ethanol with methanol, xylene, toluene and styrene exposures; smoking with exposure to *o*-cresol; and phenol with exposure to benzene.

Such mixed exposures and the associated potential for biological interactions make the health surveillance of laboratory personnel, using BM and BEM, difficult.

2.3 Adverse Effects on Health

As has been suggested previously, in a narrow scientific sense the term 'Health Surveillance' could be restricted to the monitoring of adverse effects on health and would therefore exclude BM and BEM. However, in common usage, and as implied by legislation, 'Health Surveillance' is used in a wider interpretation to encompass biological monitoring and biological effect monitoring as well as the monitoring of specific effects on health. Thus the Control of Substances Hazardous to Health Regulations in the UK embrace the wider version of 'Health Surveillance', and the European Community approach is similar.

To further complicate the issue, some people use the term 'Health Surveillance' very loosely to include assessments of fitness for work, *e.g.* tests of eyesight or of general health which some employers arrange on a regular or *ad hoc* basis. For the purpose of this chapter 'Health Surveillance' does not cover these issues but limits itself to those aspects relevant to the exposure to dangerous substances in the workplace.

3 When is Health Surveillance Needed?

Broadly speaking health surveillance is needed when the risks warrant it and when the appropriate techniques are available. Therefore in the first instance for an informed judgement to be reached a clear distinction has to be drawn between 'hazard' and 'risk'. 'Hazard' is the potential to cause harm. 'Risk' is the likelihood of harm in clearly defined circumstances (of exposure). It therefore follows that no amount of perusal of safety data sheets or information about hazards is likely to be adequate to determine whether risks are high enough to warrant health surveillance in a particular workplace with its own specific work practices and hence exposures. Similarly, studies of epidemiological literature showing an increased risk of say cancer, in a very heterogenous group of workers, does not necessarily mean that there is an increased risk in any specific subset of workers unless data from that subset are carefully studied separately.

The difficulty in interpreting epidemiological data is illustrated in a report on relative cancer risks in laboratory workers in England and Wales.[7] The occupational groups were diverse, and ranged from soil chemists, to pure mathematicians, meteorologists, ergonomists, bacteriologists, and pharmacologists. Whilst the study, in general, did not show any unusual patterns of cancer other than perhaps a small excess of brain and nervous system cancer, had there been a significant excess of any tumours, any attempts to speculate on the causative factor would have been extremely difficult with such broadly defined categories or job groups. Furthermore, given that cancers associated with occupation have a latent interval, then the difficulty would be further exaggerated in trying to trace retrospectively the chemicals and substances that might have been involved. Similar difficulties can arise in relation to the interpretation of associations between work in laboratories and an apparent increased risk of other detrimental health effects such as an adverse pregnancy outcome.

Recently, a substantive epidemiological study organized by The Royal Society of Chemistry has been published.[8] Not surprisingly it shows a reduction in overall age adjusted mortality rate amongst nearly 15,000 male professional chemists who were members of the Society. This is to be expected from their social status and is termed the 'healthy worker effect'.* Excess mortality from certain tumours of the lymphatic system and blood forming organs and some other sites was found as well as an excess mortality from some mental and neurological disorders. While these data could be the consequence of inadequate control of exposure to various organic solvents, they are not derived in such a way as to permit decisions to be made on health surveillance requirements in specific laboratory workers.

Thus for practical purposes there are three criteria each of which must be fulfilled in turn for health surveillance to be justified.

i)   *There has to be a hazard.* An adverse health effect associated with the hazardous substances should have been recognized. The basis for this evidence would ideally be from human data, but animal toxicity studies can help. One special problem faced by chemical laboratories and pilot plants is that they may be dealing with novel chemical entities about which there may be little if any animal toxicity data and probably none that are relevant to humans. In these circumstances the health hazard might be inferred by analogy with substances of similar molecular structure.

ii)   *The health risk has to be high enough.* An assessment of the risks of adverse health effects in the particular workplace should show that the likelihood of an adverse effect is appreciable enough to warrant health surveillance. Any well-stocked chemical laboratory could easily contain such a wide range of hazardous substances that a high enough exposure could harm target organs. However, a careful review of the frequency, quantity and mode of use of these substances would hopefully reveal very few circumstances in which adverse health effects were likely enough to warrant health surveillance.

---

*People in employment tend to be of a higher social class and healthier than the unemployed. Healthier people are more likely to find jobs and to remain in employment. These trends might mask subtle adverse effects of work on health.

iii) *It should be easily measurable.* There should be valid techniques for detecting the adverse health effect. Essentially there are two strands to this requirement — scientific criteria of sensitivity and specificity, and ease of interpretation (see below). The techniques should also be suitable in terms of their safety for and acceptability by the workforce.

4 Who Should Have Health Surveillance?

When one attempts to design a health surveillance programme for a specific occupational group such as laboratory staff, it is essential to have a detailed knowledge of the numerous tasks that exist in the jobs where there might be potential for exposure, and a familiarity with the catalogue of substances, mixtures and single agents to which potential exposure can occur. Such information is fundamental if one is to consider introducing biological monitoring or biological effect monitoring, or to monitor adverse health effects.

Since laboratory workers are a diverse group, it is important to qualify the job description. For each and every managerial sphere of responsibility, the work, associated tasks, and hence potential exposures depend on the type of industry, process or laboratory under consideration. Moreover, even within a specific department there may be individuals with specific bench tasks, say in weighing-out agents of high biological potency, in solvent-extraction procedures, or involved in pilot-manufacturing processes, who are effectively in a job category of their own. Observation of the tasks of these individuals as they carry them out, assisted perhaps by occupational hygiene measures of personal exposure, can help an informed judgement to be made. Useful information can also be obtained from past records of accidents and incidents, or else from reports of ill health such as dermatitis/eczema or work related respiratory symptoms.

Example 1 illustrates the need for health surveillance for laboratory workers with specific bench tasks who were vulnerable to exposure from a variety of sources.

If there is doubt it is as well to play safe and include rather than exclude employees at least for the first round of health surveillance. After the results of the health surveillance for the group of employees in question, together with a review of the occupational hygiene assessment, it may then be possible to define more clearly those employees who need continuing surveillance.

Example 1

A laboratory technician developed an itchy skin rash and wheeze, and the timing of the relevant symptoms was consistent with an occupational cause. Assessment of the laboratory complex revealed a wide range of significant exposures relevant to the index case and other co-workers. These included aldehydes (formaldehyde and glutaraldehyde), tryptic enzymes and beta-lactam antibiotics. There was a clear need to undertake health surveillance because of the risk of sensitization of the skin (urticaria, dermatitis), eyes (conjunctivitis), and respiratory tract (rhinitis and asthma).

A programme of systematic (COSHH) health risk assessments was embarked upon together with control measures. Those employees who remained significantly exposed (*e.g.* in weighing operations, making up solutions, cleaning, *etc.*) were deemed to require continuing health surveillance. Essentially, this consisted of a regular programme of, at least yearly, completion of a questionnaire, skin inspection and lung function testing. In some of the higher risk subsets the health surveillance was more frequent and rigorous.

---

5 What Health Surveillance Techniques are Appropriate?

5.1 Scientific and Ethical Criteria

Health surveillance techniques should fulfil certain scientific and ethical criteria (Table 2). From the scientific view it is obvious that the technique should be able to identify the health effect, or the uptake of a hazardous substance in the large majority of cases where these effects or levels are significant. Yet at the same time the technique must attribute the health effect, or the measured chemical substance in the body, to the chemical hazard in the workplace with measurable specificity. Thus for example, it is well known that high exposure to benzene can damage the bone marrow and cause reduced production of blood cells and hence anaemia. However, blood tests for anaemia would not be a sensitive means of health surveillance for benzene exposure since uptake would have to be unacceptably high in a large number of people for a long period of time before anaemia was evident in even a proportion of them — by which time it would be too late to behave responsibly towards any of the workers. Moreover, anaemia can be a manifestation of other ill health entirely unrelated to benzene exposure.

Table 2 *Features of health surveillance techniques*

Scientific

Sensitivity:	Measure of the probability of correctly identifying uptake or adverse effect when it is present
Specificity:	Measure of the probability of a negative result correctly identifying the absence of uptake or an adverse effect

Ease of interpretation

Ethical

Safety

Acceptability

5.2 Choice of Health Surveillance Techniques

The selection of health surveillance techniques relevant to laboratory staff depends on the hazardous nature of the substances handled, the health risks associated with these substances, and the valid techniques available for biological monitoring, biological effect monitoring, and the identification of adverse effects on health. These techniques can be classified and presented according to the hazard of the chemical or other agents, the adverse health effects, or the tests used. Each of these classifications is valid and justifiable. However, the approach that will be used here is one that best illustrates the earliest prospects of prevention and intervention.

5.3 Examples of Biological Monitoring and Biological Effect Monitoring in Health Surveillance

In practice, only a few tests based on BM or BEM can be routinely used to monitor exposure to chemicals. The following criteria should be considered before introducing such techniques:

• The parameter to be analysed must be sufficiently specific and have adequate sensitivity;

- Analytical and biological variability of the test must be acceptable;

- The test should be non-invasive to the individual;

- There should be a benefit to the individual undergoing the tests in terms of health risk evaluation.

For example, aliphatic halogenated hydrocarbons are metabolized by oxidation to the corresponding alcohols, aldehydes and acids which are excreted in the urine. Thus a suitable form of health surveillance for workers significantly exposed to 1,1,1-trichloroethane could consist of measuring urinary trichloroacetic acid at the end of the working week. n-Hexane can cause serious nerve damage if exposure is high enough. It is metabolized by oxidation of the subterminal (ω-1) carbon atoms to alcohols and then ketones. Measurement of urinary 2,5-hexanedione can constitute a useful urinary form of health surveillance through biological monitoring. Understanding the principles of the metabolism of solvents, together with detailed knowledge of work practices in their use, should help judgements based on appropriate biological monitoring[9,10] find a place in health surveillance in the future, after further refinement and validation.

Example 2 illustrates how BM in conjunction with an assessment of work practices can lead to improvements in laboratory hygiene.

The biological monitoring technique of measuring xenobiotic interactions with DNA and other macromolecules can be used as an indicator of exposure.[11] Other techniques for BM include cytogenetic studies[11] (also described in Chapter 7). Nevertheless, whilst the sensitivity, and to a lesser extent specificity at a molecular level is remarkable, the predictive value of these techniques in terms of health risk has not been established. They are also prone to some of the same variables as the more 'classical' biological monitoring techniques.

Research work with benzene illustrates some of these points in terms of the variability and interaction of adduct formation, associated with the various metabolites.[12,13] Some recent work on the cytogenetic analysis of secondary leukaemias has shown that chemically induced leukaemia such as the acute myeloid leukaemia seen following high levels of benzene exposure may have a different mechanism to primary leukaemia, and further work may identify a different and specific mechanism.

In the last few years, work has been carried out on the activation of cancer associated genes following occupational exposure to chemicals.[14] These have mainly been genes associated with lung malignancy.[15,16] Such molecular biology techniques may improve our understanding of the mechanism of chemical carcinogenesis and might provide a means, in the future, of monitoring individuals susceptible to specific disease processes.

Example 2

The laboratory of a firm involved in a major motorway project used copious quantities of dichloromethane (methylene chloride) in an extraction-based assay of the asphalt/tar used to coat the aggregate laid on the road surface. An occupational hygiene assessment revealed poor laboratory practice and inadequate local exhaust ventilation and other control measures. In parallel with advice about improvements in control and practice, a health surveillance exercise was carried out. This involved measurement of dichloromethane and its metabolite carboxyhaemoglobin (COHb) from a blood sample taken from technicians at the end of an 8-hour working shift. The results confirmed a significant uptake of dichloromethane (probably from both skin absorption and inhalation). After implementation of control measures a repeat health surveillance exercise revealed no significant uptake of dichloromethane and hence this surveillance was not repeated although the introduced rigid control measures remained in force.

5.4 Surveillance for Symptoms of Ill-Health

Many lay people, and indeed some scientists, are under the misapprehension that adverse effects can commonly be revealed at a presymptomatic stage. This is often not the case and awareness of the relevant symptoms following exposure is an important part of health surveillance.

5.4.1 *Symptoms of Dermatitis/Eczema*[17]

Exposure to defatting agents can result in irritant dermatitis with symptoms such as rashes, dry skin, or scaling on the hands. Other hazardous agents can cause dermatitis or eczema through sensitization. These range from certain compounds of transition metals such as nickel, cobalt and chromium, to di- and polyamines, various alcohols, aldehydes, and several hundred proven distinct chemical agents. Other rarer skin disorders vary from pigmentation to special forms of acne and, very occasionally, cancer.

5.4.2 *Symptoms of Conjunctivitis, Rhinitis or Asthma*[18,19]

Many substances and practices in laboratories or in pilot plants can cause respiratory symptoms suggestive of inflammation in the airways of the nose or lungs, for example exposure to di-isocyanates, diamines, aromatic acid anhydrides, certain aldehydes such as formaldehyde or glutaraldehyde, antibiotics, and laboratory animals.

Even devices used for personal protection, notably rubber gloves, can cause asthma (as well as dermatitis).

Questionnaires can address the temporal pattern of the symptoms beside the nature of the symptoms themselves and can therefore be used with a high degree of specificity for the purposes of health surveillance, if properly designed.[20]

5.5 Tests for Adverse Health Effects

Several tests of body function can lend themselves with varying degrees of specificity and sensitivity for health surveillance purposes. Thus tests of respiratory function, notably of forced airflow on expiration, can, if repeated on an appropriate timescale, be used to help diagnose occupational asthma.[19] In a mineralogy laboratory where high noise exposures could arise from grinding tools, pure tone audiometry can be used to detect noise-induced hearing loss (Example 3).

Example 3

A mineralogy laboratory of a petrochemical company employed geologists and technicians on a variety of tasks. These included the grinding of silica-bearing and other rocks into fine sections for microscopical study. A risk assessment of the workplace revealed high (>90dBA) levels of noise exposure associated with grinding operations. Moreover grinding of samples resulted in high personal exposure (>0.4 mg m^{-3}) to respirable silica (quartz). The control strategy that was implemented, besides better segregation, personal protection, *etc.* also included health surveillance. This included regular pure tone audiometry to assess noise-induced hearing loss and chest radiography for signs of pneumoconiosis (notably silicosis).

Some tests rely on identifying structural rather than functional abnormalities. Thus chest X-ray films can show the abnormal appearances of specific pathological changes following a high enough exposure to certain mineral dusts such as crystalline silica.[19,21] Abnormal cells can be identified microscopically in the urine of workers exposed to those aromatic amines which can cause bladder cancer.[22]

5.6 Health Records and Prospective Studies of Work-Styles and Lifestyles

Adverse pregnancy outcomes, such as miscarriages or congenital abnormalities, and cancers are more common than the lay person believes them to be. It is the exception rather than the rule that when one of these unwanted events occurs, it can be reliably ascribed to occupational exposure in any individual case. With cancers in particular there can be a long time lag (latency) (up to 40 years) before exposure to a hazardous substance results in symptoms of the disease.[22]

Health surveillance for reproductive or carcinogenic effects can take the form of keeping an accurate prospective record. Arguably this might not be of much physical benefit to the individual worker who suffers such a serious effect on health, but could help in compensation claims and by identifying risks which would not be recognized without a large enough body of statistical data. Thus some employers encourage their female employees to report the onset of pregnancy and prospectively log the outcome.

For health surveillance in relation to suspect carcinogens, employers must record an adequately detailed occupational history of their employees together with personal identifiers including National Insurance Numbers, and retain this information for at least 40 years.

If common procedures are adopted throughout chemical laboratories to identify, report, and record ill-health from work-related disease as recommended at a previous Royal Society of Chemistry symposium on laboratory safety,[23] and if these data are pooled and suitably analysed, valuable information may result to guide risk assessments and decisions about future health surveillance. The Royal Society of Chemistry's Effect of Chemicals Assessment Programme (RECAP), described in Chapter 1, is one attempt to monitor the work-style and lifestyle of professional chemists with the aim of generating such data.

6 Implementation of Health Surveillance

6.1 Who Should Carry Out Health Surveillance?

First and foremost, employees must be made aware by their employer of the risks to health of the hazardous substances to which they are exposed. Besides being a statutory legal requirement, risk assessment greatly facilitates the implementation of health surveillance. Once employees are aware of the health risks they are more likely to accept health surveillance enthusiastically. Furthermore, employees themselves can contribute to the first line of health surveillance by reporting symptoms such as those of dermatitis or asthma.

Managers or other 'responsible persons' without any formal health training can also contribute to health surveillance, for example, by inspecting hands on a regular basis and then referring any employee whose skin looks in any way abnormal to the occupational health physician or occupational health nurse. Thus a pyramid approach can be used where a comprehensive strategy is implemented on the advice of a specialist occupational physician, and then its day-to-day implementation can rely on the employees themselves, their managers or other health professionals, with specialist assessment being required in only a minority of employees (Figure 2).

Figure 2 *Hierarchy of health surveillance approaches*

The COSHH regulations[3] lay down specified circumstances of exposure for which health surveillance is not only compulsory but must be carried out by a doctor employed by the Health and Safety Executive (an Employment Medical Adviser) or a doctor appointed to act on their behalf — and who could be the firm's occupational physician.

6.2 What Should Be Done with the Results, and about Them?

In the first instance the individual employee must be notified of the results of his/her health surveillance and guided as to their interpretation. Results must also be collated in a summary form, usually by the occupational health department, in a manner which does not identify individuals. The law requires that these summary results are brought to the attention of the relevant employee and their representative. The Safety Committee is a good forum for conveying this information.

When adverse health effects are identified, or when adverse trends are noticed in biological monitoring, steps must be taken to reassess the working environment and reduce exposure. Often the erroneous supposition is made that some peculiar individual susceptibility is responsible. This is rarely the case. Even when only one individual manifests an adverse effect he/she should be looked upon as the sentinel case or the 'canary' and other sufferers may follow unless exposure is controlled. With appropriate timely steps the workplace relocation of the affected employee may be avoidable, but sometimes when sensitization to very low concentrations results in severe symptoms of asthma or of generalized allergy, relocation may be the only option.

6.3 Repetition of Health Surveillance and Record Keeping

The frequency with which health surveillance is carried out depends on the characteristics of the health risk for which the surveillance is being undertaken and the outcome in any given individual. Thus, if a new employee starts work in an environment which poses an appreciable risk of occupational asthma, besides a baseline assessment of symptoms and lung function on commencement of employment, health surveillance might need to be repeated after three months, a further six months and yearly thereafter if all appears well.

If dermatitis is the subject of health surveillance then, depending on the frequency with which this occurs, health surveillance could range from a fortnightly inspection by a competent person to a very simple self-completed yearly questionnaire.

Access to health records is a very important issue. Employees have a right of access to all their health records, on request, according to The Access to Health Records Act, 1990. Employers only have a right, when the health surveillance is statutory, to a simple statement of outcome for each individual, usually on a 'fit/unfit' basis. Other data can only be accessed by employers in descriptive summaries of pooled data from their employees.

By law under the COSHH Regulations,[3] Health Surveillance records must be kept for at least 40 years from the last entry therein.

7 Conclusion

Health surveillance is one part of a comprehensive strategy for working with dangerous substances in the laboratory environment. While it is never the mainstay of managing hazardous exposures, health surveillance can yield valuable information for protecting the individual and improving control of the working environment for all concerned.

8 References

1.   A.W. Musk, J.M. Peters, and D.H. Wegman, *Am. J. Ind. Med.*, 1988, **13**, 331.

2.   X. Baur, *Lung*, 1990, **168**, (Supplement), 606.

3.   Health and Safety Commission, 'Control of Substances Hazardous to Health and Control of Carcinogenic Substances, Control of Substances Hazardous to Health Regulations 1988', Approved Codes of Practice, 4th Edition, HMSO, London, 1993.

4. Health and Safety Executive, 'Surveillance of People Exposed to Health Risks at Work', HMSO, London, 1990.

5. 'Casarett and Doull's Toxicology: The Basic Science of Poisons', ed. M.O. Amdur, J. Doull, and C.D. Klaasen, 4th edition, McGraw Hill, New York, 1991.

6. P. Grandjean, 'Skin Penetration: Hazardous Chemicals at Work', Taylor & Francis, London, 1990.

7. L. Carpenter, V. Beral, E. Roman, A.J. Swerdlow, and G. Davies, *Lancet*, 1991, 1080.

8. W.J. Hunter, B.A. Henman, D.M. Bartlett, and I.P. Le Geyt, *Am. J. Ind. Med.*, 1993, **23**, 615.

9. D. Gompertz, *Ann. Occup. Hyg.*, 1980, **23**, 405.

10. Health and Safety Executive, 'Biological Monitoring for Chemical Exposures in the Workplace', Guidance Note EH56, HMSO, London, 1992.

11. 'Human Carcinogen Exposure: Biomonitoring and Risk Assessment', ed. R.C. Garner, P.B. Farmer, G.T. Steel, and A.S. Wright, Oxford University Press, Oxford, 1991.

12. G.F. Kalf, *Crit. Rev. Toxicol.*, 1987, **18**, 141.

13. A. Yardley-Jones, D. Anderson, D.V. Parke, *Br. J. Ind. Med.*, 1991, **48**, 437 and references therein.

14. P.W. Brandt-Rauf, S. Smith, H.L. Niman, M.D. Goldstein, and E. Favata, *J. Soc. Occup. Med.*, 1989, **39**, 141.

15. P.W. Brandt-Rauf and H.L. Niman, *Br. J. Ind. Med.*, 1988, **45**, 689.

16. P.W. Brandt-Rauf, S. Smith, and F.P. Perera, *et al.*, *J. Soc. Occup. Med.*, 1990, **40**, 11.

17. 'Essentials of Industrial Dermatology', ed. W.A.D. Griffiths and D.S. Wilkinson, Blackwell Scientific Publications, Oxford, 1985.

18. R.M. Agius, J. Nee, B. McGovern, and A. Robertson, *Ann. Occ. Hyg.*, 1991, **35**, 129.

19. W.K.C. Morgan and A. Seaton, 'Occupational Lung Diseases', 2nd Edition, Saunders, Philadelphia, 1984.

20.   A. Seaton, R. Agius, E. McCloy, and D. D'Auria, 'Practical Occupational Medicine', Chapter 3, Edward Arnold, London, 1993.

21.   International Agency for Research on Cancer, 'Silica and Some Silicates', IARC Monographs on the Evaluation of the Carcinogenic Risk of Chemicals to Humans, Volume 42, IARC, Lyon, 1987.

22.   'Hunter's Diseases of Occupations', ed. P.A.B. Raffle, W.R. Lee, R.I. McCallum, and R. Murray, Edward Arnold, 1987, Chapter 24.

23.   'Health and Safety in the Chemical Laboratory — Where do we go from here?', Royal Society of Chemistry, London, 1984.

CHAPTER 3

An Insurer's Approach to Health, Safety and Environmental Issues in the Laboratory

I. WRIGHTSON

1 Introduction

When an employer fails in his duty of care towards his employees and they suffer injury or disease, those employees are entitled to sue their employer for damages to compensate them for the injury or disease which they have suffered. In order to ensure that sufficient funds are available for such compensation, the Employers' Liability (Compulsory Insurance) Act 1969 requires all employers (with one or two exceptions such as nationalized undertakings) to take out an insurance policy to cover their legal liability to their employees.

The insurer provides cover for the employer under the Employers' Liability policy. The annual premium of this policy is based on a percentage of the wages paid to employees and takes into account the type of industry, the risks and the claims experience. The Insurer expects the Insured to take appropriate action to prevent accidents and occupational ill-health and to provide good standards of health and safety. This can be done by the introduction of a risk management system to prevent (or minimize) claims of negligence by his employees. If a claim arises the Insurer looks to the Insured to provide evidence to refute the claim. Such evidence should be capable of withstanding independent examination. A systems approach to risk control provides an adequate framework to allow such evidence to be available. By using this framework the Insurer and the Insured can work together so that both can remain commercially viable.

2 The Cost of Insurance

In recent years the cost of Employers' Liability claims has risen dramatically and this, along with several other factors, has led to increases in the cost of insurance. There have been sharp increases in claims arising from accidents and industrial disease over the last five years (1987-1992) resulting from a greater awareness of the right of individuals to compensation. Claims for accidents have risen by 33% and claims for occupational deafness by 186%. Apart from the well documented diseases such as asbestosis and occupational deafness, there are growing numbers of claims for byssinosis, a disease arising from the cotton textile industry, and occupationally induced cancers. Repetitive strain injuries are increasing particularly amongst clerical workers, such as keyboard operators.

The time lag between actual exposure to harmful chemicals, fibres or gases, and manifestation of disease may be anything up to fifty years. For that reason it is not possible to forecast with any accuracy what the potential liabilities will be in the future.

Inflation has a major effect on claim settlement costs. For reasons outside the control of insurers it can take many years to settle some Employers' Liability claims. This is especially significant in a case of severe injury which is not only extremely costly, but medical prognosis may not be known for a number of years. Today's premiums must provide for the settlement of these claims and take into account the fact that the eventual settlement, perhaps five or six years later, will be based on the levels which are applicable at that date. It must also be borne in mind that legal costs and experts' fees continue to rise. Even a successful defence may mean that an Insurer does not cover his own legal costs, particularly where the claimant has no money and is legally aided.

3 The Cost of Accidents

The cost of accidents may be divided simply into those which can be covered by insurance and those which are not insurable.

The insurable costs are predominantly covered by the Employers' Liability policy which, as indicated above, is compulsory and a Public Liability policy which is voluntary. The Public Liability policy provides cover for the company's liability to members of the public who are not their employees for any accident, injury, or damage to them or their property.

When an accident occurs which causes injury to an employee a number of costs arise which cannot be offset by insurance. Examples of such costs are as follows:

- Cost of injured person's lost work time;

- Cost of wages paid to persons who go to the assistance of the injured person;

- Cost of wages paid to persons who stop work out of curiosity or sympathy;

- Cost of wages for persons who were unable to continue work because they relied on injured person's aid or output;

- Cost of damage to material;

- Cost of damage to plant;

- Cost of supervisor's time spent in assisting, investigating, reporting, reassessing work, and making necessary adjustments;

- Cost of instructing a replacement worker;

- Cost of first aid or medical facilities;

- Cost of time spent by administration staff in processing investigations and reports;

- Cost of any action taken by the Enforcing Authority; and

- Cost of any adverse publicity.

The costs of accidents[1] have been studied recently by the Health and Safety Executive in five organizations over a period of 13 to 18 weeks. Only a small proportion of the costs were covered by insurance. The un-insured costs were between 8 and 36 times greater than the cost of the insurance premium paid during the period of study. In one case the costs of accidents represented 37% of the annual profit, in a second case they were 8.5% of the income, and in a third case they were 5% of the operating costs. During the limited period of study, no accident occurred in any of the five firms involving a death, major injury or large scale loss due to fire or explosion. There were no civil claims or prosecutions to swell the basic figures.

4 Viewpoints of Interested Parties

The employer, the employee, and the Insurer have different views on the general approach to health and safety. It is important to take these viewpoints into consideration before outlining how health and safety should be managed.

Often, the employer believes that:

- The risks are known;

- His staff understand their health and safety responsibilities and can perform them effectively, with little or no training of line management;

- The extent of risk control is measured in accident severity and by the accident record;

- Health and safety is a separate entity, divorced from commercial success;

- Compliance means keeping just within statute law with little or no knowledge of common law; and

- Employers' Liability Insurance is a compulsory overhead, largely unrelated to the operating risks, particularly those of a long-term nature.

On the other hand, the employee believes that:

- The employer is responsible for all risks at work;

- Risks can be taken on one's own initiative and assessment of a situation; and

- There is no individual responsibility to others at work for one's acts or omissions.

Finally, the Insurer believes that:

- Management can be well intentioned but there is often a gap between intent and accomplishment;

- Few managements understand common law obligations and how to demonstrate discharge of duty;

- The insurance profession's experience of servicing the needs of clients has provided an ability to understand the practicalities of problem-solving, and knowledge of the best systems for generating the information critical for performance assessment;

- That same experience gained across the entire spectrum of commercial activities enables fundamental deficiencies to be recognized in risk identification, the setting of objectives, definition of responsibilities, existing control systems, communications (both internal and external), problem solving and performance assessment; and

- Health and safety is an integral part of the business activity and commercial success, enabling the crucial resource — the employees — to make a maximum contribution because they feel that they matter as individuals, they know what is expected of them and they are kept informed.

In order to ensure that good standards of health and safety are achieved and that positive action is taken to reduce accidents and prevent claims, the Insurer advocates a pro-active systematic approach to health and safety. Such an approach involves the reconciliation of the viewpoints outlined above and active co-operation between the employer, the employee and the Insurer.

5 A Systems Approach to Risk Management

Eagle Star is one of the leading Employers' Liability Insurers in the UK. In order to support this role, Eagle Star pioneered health and safety risk management appraisals in 1943. In 1988, Eagle Star introduced a new approach to the management of health and safety, the objective being to prevent harm by using a systems approach to identify risk and then to take appropriate action to eliminate, reduce or control that risk. These systems can then be monitored and evaluated to assess their effectiveness.

This approach was based on various Health and Safety Executive publications (relating to safety policies,[2] managing safety,[3] monitoring safety,[4] and accident analysis[5]), the principles of common law, and quality assurance procedures (BS 5750[6]).

In recent years quality has been promoted as a major factor for commercial success. It involves a systems approach in which everyone is motivated to strive for improvements. In a broader concept, this approach can be applied to any potential loss of resources or profit, including loss of employee morale or time. Poor health and safety conditions can lead to accidents and losses, as demonstrated earlier, and the analogy with quality assurance can be drawn. Health and safety must be seen as an integral part of an organization's success. It is essential that all parties co-operate and work together to prevent these losses.

The Insurer advocates that the Insured applies the requirements of both the Statute and Common Law using a quality assurance type approach, incorporating proper monitoring and record keeping, in order to reduce the number and severity of accidents and the incidence of occupational disease. This approach will lead to cost savings, improved by safety records, increased productivity, increased prestige, and increased profitability. In 1991, the Health and Safety Executive also drew attention to the similarities between their approach to good health and safety management and the principles of total quality management.[7]

When examining the health and safety performance of the Insured, the Insurer applies the principles of both the Statute and the Common Law.

The quality and nature of the Insured's health and safety performance are evaluated in terms of the Statute Law using the criteria advocated by the Health and Safety Executive:[3]

- The elimination of hazards by compliance with legislation and codes of practice;

- The operation of the health and safety policy;

- The accident and ill-health records; and

- The progress towards long term objectives.

The evaluation of the Insured's Common Law duty of care for his employees is undertaken in terms of the following simplified criteria:

- Compliance with statute law;

- The identification of foreseeable risks;

- The application of relevant risk controls; and

- Ensuring the controls are effective.

6   The Needs of the Employer

In order to set up a quality assurance type of approach to the management of health and safety along the lines advocated above the Insured needs to have four key systems in place:

- A regime of gathering information to enable risks to be identified;

- A method of assessing risks and formulating acceptable risk controls;

- A programmed evaluation of risk controls; and

- A procedure for the investigation and correction of deficiencies.

For self-regulation an employer needs a management framework that will cope with all aspects of the commercial operations, as well as health and safety. This framework must be capable of generating commitment by the whole workforce to the success of whatever aspect of the business is addressed. In order to ensure that these key systems for the management of health and safety are established the Insured must provide the following:

- A positive policy on the assessment and control of risks, involving the participation of both management (at all levels) and employees in all systems;

- A progressive risk reduction programme;

- Safe systems of work and operating procedures;

- Agreed disciplinary procedures;

- Periodic safety inspections/audits;

- Investigations of all accidents, dangerous occurrences, near misses, and cases of occupational ill-health;

- Annual evaluation of health and safety performance and the setting of future objectives by senior management;

- Clear records to demonstrate achievements on all the above matters including decision-making processes, with permanent retention; and

- Periodic liaison with the Insurer to update the state of knowledge.

The employer needs to be able to demonstrate that he has discharged his duties and obligations with respect to health and safety to both Insurer and the Enforcing Authority and more importantly, and ultimately, to the Courts.

The employers' duty of care to his employees was clearly defined by Mr. Justice Swanwick in the case of Stokes v Guest, Keen and Nettlefold (Bolts and Nuts) Limited,[8] as:

'... The overall test is still the conduct of the reasonable and prudent employer, taking positive thought for the safety of his workers in the light of what he knows or ought to know; where there is a recognized and general practice which has been followed for a substantial period in similar circumstances without mishap, he is entitled to follow it, unless in the light of common sense or knowledge it is clearly bad; but, where there is developing knowledge, he must keep reasonably abreast of it and not be too slow to apply it; and where he has in fact greater than average knowledge of the risks, he may be thereby obliged to take more than the average or standard precautions. He must weigh up the risk in terms of the likelihood of injury occurring and the potential consequences if it does; and he must balance against this the probable effectiveness of the precautions that can be taken to meet it and the expense and inconvenience they involve. If he is found to have fallen below the standard to be properly expected of a reasonable and prudent employer in these respects, he is negligent.'

Conversely, if the employer has taken all reasonably practicable measures which were appropriate at the material time, he should not be held negligent. It must be emphasized, however, that the employers' duty of care is a very onerous one.

7 The Health and Safety Policy

The provision of a written health and safety policy by the Insured is the basic requirement for establishing good management systems for the control of health and safety. The policy should state the Insured's intentions and approach to health and safety and should give details of the organization and arrangements for its effective implementation. The policy is the key to high standards of health and safety in the laboratory.

The Insurer needs to establish that the statement of intent which introduces the policy demonstrates the Insured's commitment to health and safety at the highest level within his organization. To this end the policy should be signed and dated by the Chief Executive. The content of the policy should be effectively communicated to all employees and a record should be kept of this action.

The policy should be a living document which should be reviewed at least annually, preferably when the health and safety performance is evaluated, and should be revised when appropriate.

The policy should take into account the requirements of the Management of Health and Safety at Work Regulations 1992[9] in terms of risk assessments and health and safety arrangements. Other legislation which is directly relevant to the laboratory such as that relating to the control of substances hazardous to health[10] and environmental issues should also be taken into account.

In the organization section of the policy, the Insurer would expect to see details of the duties and responsibilities of each successive level of line management, and those competent persons appointed by the Insured to provide health and safety assistance, such as the Laboratory Safety Officer. The inclusion of an organization diagram within the policy would demonstrate the inter-relationship between the relevant individuals. The duties and responsibilities of employees should also be included along with information about safety representatives and the organization, role and constitution of the Laboratory Safety Committee.

In the arrangements section of the policy the Insurer would expect to see general information about risk assessments, training, safe systems of work and working procedures, accident reporting and investigation, first aid, fire safety and other emergency procedures, the control of contractors and other visitors and the laboratory safety rules. Detailed procedures[11] arising from the policy need to be drawn together in the form of a Laboratory Safety Manual which should include information on the following:

- The storage and use of hazardous substances (*e.g.* compressed gases, cryogenic liquids, flammable liquids, toxics, corrosives, irritants, radioactive chemicals);

- Storage and handling of micro-organisms;

- Hazardous procedures (vacuum, high pressure, high temperature, ionizing radiations, non-ionizing radiations);

- Operating instructions for specific items of plant and equipment;

- Safe use and maintenance of electrical equipment;

- The provision of effective control measures to prevent exposure to hazardous substances, including suitable personal protective equipment;

- The regular examination of equipment such as pressure vessels, lifting tackle, fume cupboards, and other local exhaust ventilation;

- The safety of maintenance staff and the safety of employees who may be affected by maintenance work;

- Arrangements for out of hours working;

- Unattended operations; and

- The disposal of waste materials.

The manual should also take into account everyday activities such as lifting, carrying and other manual handling and the handling of glassware, as well as more specialized procedures specific to particular laboratories. The procedures contained in the manual should be based on the risk assessments undertaken by the Insured.

8 Risk Assessments

On 1st January 1993 six new Codes of Health and Safety Regulations became law. These Regulations apply to the management of health and safety at work,[9] manual handling operations,[12] display screen equipment,[13] personal protective equipment,[14] the workplace,[15] and work equipment.[16] These Regulations update the Health and Safety at Work etc. Act 1974 and place specific duties on both employers and employees. The Regulations repeal old and outdated legislation and provide for the establishment of up to date standards at all workplaces. A new Code of Fire Regulations for Workplace Safety[17] is expected to be made during 1994.

The most important of these new Regulations for laboratories is the Management of Health and Safety at Work Regulations 1992.[9] These Regulations in effect make good health and safety management a legal requirement and focus on three main areas:

- Risk assessments;

- Arrangements for health and safety, including training; and

- Employees' duties and responsibilities.

The Regulations on manual handling, display screen equipment, personal protective equipment, work equipment and the proposed Fire Regulations also require risk assessments to be undertaken.

The Insured needs to be able to demonstrate to the Insurer that he has used a systematic approach to the identification of all hazards and the assessment of risk along with the provision of appropriate control measures, including safe systems of work. The most important hazards in the laboratory may have been identified in the past, and informal risk assessments may have resulted in written safe systems of work being included in the Safety Manual. These risk assessments should now be formalized and recorded. The content of these safe systems of work need to be reviewed and, if appropriate, revised to ensure that they remain relevant and up to date. New safe systems of work may also be required. In Common Law it is important that all risks are recorded as it is often incidents arising out of relatively minor risks, such as slips, trips, cuts and bruises, which give rise to claims for compensation.

There is no specific requirement for the quantification of risk but the concept of quantification is implied in the Approved Code of Practice[18] for Management of Health and Safety at Work Regulations in order to set priorities for the introduction of control measures. The Insurer supports the idea of quantification and advocates that the Insured follows at least the simple quantification set out in the Health and Safety Executive's publication 'Successful Health and Safety Management'.[7] In this publication risk is defined as the product of hazard severity and the likelihood of occurrence. Numerical values of 1, 2, and 3 are used for the increasing hazard severity and the increasing likelihood of occurrence. Using these values in the formula:

$$Risk = Hazard\ Severity \times Likelihood\ of\ Occurrence$$

numerical values of risk from 1 to 9 can be generated and used for setting priorities in relation to the provision of control measures. Numerical values of 1 to 6 have been advocated elsewhere[19] for increasing hazard severity and increasing likelihood of occurrence, giving risk values between 1 and 36. After control measures have been introduced risk values can be recalculated to demonstrate the risk improvements which have been made. Consideration can then be given as to whether any further control measures are necessary.

In the laboratory all risks arising from physical, chemical and, if appropriate, biological agencies and all work activities need to be assessed and recorded. Assessments made under the requirements of other Regulations need not be repeated but appropriate cross referencing is essential to demonstrate a comprehensive approach. Risk arising from work activities involving substances hazardous to health should have already been made under the requirements of the Control of Substances Hazardous to Health Regulations (COSHH) 1988.[10]

9 Liabilities and COSHH

The Insured who complies with the requirements of the COSHH Regulations to the satisfaction of the Enforcing Authority might still be unsuccessful in resisting relevant claims of negligence. The Insured needs to satisfy the needs of the Common Law duties and should be able to demonstrate that he is a reasonable and prudent employer.

When there is a claim of negligence in relation to COSHH the Insured needs to be able to provide evidence to the contrary. The evidence should demonstrate the efforts made by the employer, commensurate with commercial viability, to discover and control foreseeable risks. Such effort should account for the abilities of employees and the special nature of laboratory activities.

Disease-related claims can arise many years after any alleged initial exposure and often after several changes of management. The Insurer requires evidence that all the above requirements were complied with by the reasonable and prudent employer, who:

• Took note of all official, national, trade and professional guidance[20,21] and the scientific literature;

- Considered the varying abilities and vagaries of his employees in the circumstances that might arise at work; and

- Recognized the limitations of internal expertise.

The documented and systematic approach advocated by the Insurer should generate all the evidence required. It remains largely independent of changes in personnel provided that it is regularly reviewed and updated in order for it to remain effective. It is essential that employees are fully involved in this approach.

Employees should be involved in the risk management process by:

- Ensuring they are informed of why the risk assessment procedure is necessary;

- Involving them in the information-gathering process;

- Consulting them in the decision-making process, so that good sense prevails in the acceptance of any residual risks;

- Having them participate in the formulation of relevant risk controls;

- Agreeing disciplinary procedures with them;

- Regular motivation and refresher training programmes; and

- Periodic feedback of findings *via* a medium that allows for comment.

The foundation stone of the prevention of claims is the initial gathering of information.

The Insurer advocates the spreading of the information gathering task amongst all levels of employees, who have relevant knowledge and experience to contribute both obvious facts and those which are often unknown to management. The ability of employees to contribute will depend on how well they are briefed and motivated. This is the start of assembling the systematic evidence required to demonstrate discharge of duty. Records are required for this first step and each successive stage of the assessment process.

When faced with an allegation of negligence originating from a time several years previously, it is necessary for the Insurer to examine if the relevant risk was foreseeable and if it was considered. Therefore, when carrying out COSHH assessments it is important to record the step by step process of:

- Identifying all substances at all stages of a process;

- Their initial screening into general categories such as safe, domestic type risk and the remainder;

- Determining the relevant exposures of employees; and

- Determining the potential effects on persons other than employees.

It may be stating the obvious that exposures assessed as 'safe' should be recorded, but it is a common experience to find in retrospect that proving the 'obviously safe' cannot be done to the likely satisfaction of a court of law.

The COSHH Regulations require that the employer should make a suitable and sufficient assessment of the risks arising from work involving substances hazardous to health and of the steps that need to be taken to meet the requirements of the Regulations. It is essential that both aspects of this duty are fully addressed. Many employers simply, and incorrectly, make an assessment only of the potential risks arising from substances hazardous to health. Other aspects relating to the content of assessments are often ignored or forgotten.

All assessments should be fully documented, and in order to demonstrate that they are suitable and sufficient, should meet at least the specific requirements set out in the COSHH Approved Codes of Practice.[22] The assessments should be supplemented, where appropriate, with information drawn from historical findings in case law and the Employers' Liability insurance experience in relation to the ability to demonstrate the discharge of the duty of care.

When determining the exposure of employees to substances hazardous to health reference must be made to the lists of Occupational Exposure Limits set out in the Health and Safety Executive's Guidance Note EH40[23] which is revised and published annually. In laboratories many substances are used and new ones are synthesized which do not have exposure limits. The absence of any such substances from the Health and Safety Executive's list does not indicate that it is safe. As part of the assessment the employer should determine his own working practices and standards of control. In some cases, there may be sufficient scientific and technical information available to set an in-house occupational exposure limit. The setting of such standards is not a precise exercise. It calls for judgements which involve toxicological and occupational hygiene expertise. Each substance should be considered on its own merits. There is no simple formula or checklist for the derivation of such limits but some methodologies have been published in relation to pharmaceutical materials.[24-26]

COSHH assessments should be reviewed regularly and, in any case, whenever there is any reason to suspect that the assessment is no longer valid or where there has been a significant change in the work carried out. As a result of any review of an assessment, the changes which are identified should be made and recorded. The length of time between reviews depends on the nature of the risk, the work and a judgement on the likelihood of future changes. Usually reviews should be undertaken between every two to five years.

All documents generated under the COSHH Regulations should be retained in order to demonstrate the action of a reasonable and prudent employer in ensuring the health and safety of his employees. Since disease-related claims may arise at any time after exposure, and sometimes after death, the retention period for records is unlimited.

10 Environmental Issues

The systematic approach to the management of health and safety and COSHH can equally be applied to environmental issues relating to the laboratory. The Insurer expects the Insured to adopt this approach and to base his management of environmental risk on the principles set out in the British Standard BS 7750 'Environmental Management Systems'.[27]

The Insured should make an assessment of all the chemicals, processes and work activities which could have an impact on the environment. Much of the information obtained for the COSHH assessments can also be used for environmental assessments. It may be appropriate, therefore, to adopt an integrated approach to both COSHH and environmental issues by undertaking Integrated Risk Assessments.[28] As a result of such assessments steps can be taken to eliminate, reduce or control environmental risks in relation to air, water and land. The aim must always be to avoid both pollution of the environment and harm to human health.

In the laboratory, environmental issues relate mainly to the discharge of fume cupboards and the disposal of waste. However, it is important to take into account the storage of chemicals and the actions to be taken in the case of spillages or uncontrolled releases. Good housekeeping, safe storage, safe systems of work and employee training are essential prerequisites for good environmental control. The systems and procedures for the management of environmental issues can be incorporated into the health and safety policy to confirm the integrated approach.

11 Training

Training is a fundamental aspect of health, safety and environmental issues which is often not given the necessary priority. It is important that the Insured has formal policies, procedures and recording systems for training. All employees should be provided with adequate training in health, safety[9] and environmental issues.[27] Training is an important way of achieving competence and helps to convert information into safe working practices. It contributes to the organization's health, safety and environmental culture and is needed at all levels, including management.

New employees should undergo induction training which should include the arrangements for health, safety and environmental matters in terms of the health and safety policy, their duties and responsibilities, laboratory procedures and rules, first aid, accident reporting, fire safety and emergency procedures. It must not be assumed that because most new employees will have formal qualifications that they are already fully aware of health, safety and environmental issues. Academic courses generally do not include these matters. This is, however, a separate issue and needs to be addressed elsewhere. New employees may have worked in a laboratory before but they still need to be acquainted with the specific requirements of their new laboratory and their new duties and responsibilities. Particular attention should be given to the needs of younger employees. An induction checklist should be used to record this training and it should be signed and dated by both the trainer and the trainee to demonstrate that the training has been completed successfully.

Training should be an on-going feature of laboratory work and employees should be given additional training when exposed to new or increased risks arising from new equipment, new processes and new techniques. In such cases, following risk assessments new safe systems of work will need to be introduced and appropriate training will need to be given. Training will also be required when new legislation is introduced or when employees are transferred to new work or when they are given new responsibilities. Training needs should be reviewed regularly and appropriate refresher, update or motivational training should be given. All training should be fully documented and assessed to ensure that it is effective. When allocating work account should be taken of employees' abilities and capabilities and they should not be required to undertake tasks for which they are unsuitable or ill-prepared.

Training is also required by specific Regulations such as COSHH. All persons handling substances hazardous to health must be provided with appropriate information, instruction and training in order that they fully understand the requirements of the COSHH Regulations and their duties under them. Personnel who are delegated to carry out the employers' duties such as undertaking assessments should be given the necessary information, instruction and training for that task so that it can be carried out effectively. Training under COSHH should be recorded and form part of the COSHH documentation system.

The provision of health, safety and environmental training for managers and supervisors is particularly important as members of line management have special responsibilities on behalf of the Employer. All members of management should be given appropriate training to enable them to discharge their duties effectively. Such training should include the duties imposed by both statute and common law. Management's role in accident investigation and safety inspections should also be included in this training. Members of management should be aware that they must be in a position to demonstrate to an outside authority, if necessary, that:

• They have identified the hazards in their areas;

• They have implemented controls to deal with those hazards; and

• They systematically monitor the control measures to ensure that these are being effectively used.

Members of management who are appointed to functional positions such as Laboratory Safety Officer should be given special training to allow them to provide management and laboratory staff with professional advice on health, safety, and where appropriate, environmental matters.

12 Reporting and Investigation of Accidents

All accidents, incidents and near misses, no matter how trivial they may seem to be, should be recorded and investigated.

Accidents which give rise to personal injury should be recorded in the Accident Book, Form BI510[29] in order to satisfy the requirements of the Social Security Act 1975 and the Reporting of Injuries, Diseases and Dangerous Occurrences Regulations 1985[30] (RIDDOR). All entries in the Accident Book should outline how the accident occurred, the cause of the accident (if possible) and the injury sustained. Other incidents, including those relating to the environment, and near misses should be recorded in a separate book. There is a requirement under RIDDOR to report to the enforcing authority all fatal accidents, minor injury accidents, and accidents that give rise to an absence from work of more than 3 days or which renders the injured person unable to undertake normal work for more than 3 days. In addition, any of the dangerous occurrences or cases of ill-health listed in RIDDOR need to be reported. The Insurer should also be kept informed about accidents, particularly those which are likely to become claims.

The investigation of all accidents should be recorded and the report generated should give detailed information on the circumstances of the accident, the causes and the injuries sustained. The names of witnesses should be recorded and where appropriate statements and photographs should be taken. Recommendations should be made in the report on the steps which need to be taken to prevent a recurrence. The remedial action taken as a result of the investigation should also be recorded. Information arising out of investigation is important not only to prevent a recurrence but to determine whether there are wider implications for safety and to identify any new hazards which had not been foreseen so that further risk assessments can be undertaken.

The analysis of accident statistics should be undertaken on a regular basis in order to identify trends and underlying causes. Accident analysis can provide much useful information which can be used in accident prevention[5] and risk assessments and should be considered when the health, safety and environmental performance is evaluated.

13 Inspections and Audits

The laboratory's health, safety and environmental performance should be monitored by management who should carry out regular safety inspections or audits. Safety inspections should be undertaken at pre-planned intervals and should make use of a checklist which is specific to the requirements of the laboratory. Inspections are designed to ensure that safe operational procedures and safe systems of work are followed and that good standards are observed. The results of the inspections and actions arising from them should be recorded. This is a useful technique which may reveal major problem areas requiring attention especially if the reasons behind the failures are fully investigated.

Safety audits are more thorough and are a systematic critical examination of work activities in order to identify potential hazards and determine the associated levels of risk. The objective of an audit is to ensure that the organization's management systems for the control of health, safety and environment issues are operating effectively and that they are meeting the legal requirements.

Audits include an examination of the laboratory health and safety policy, organizational and operational procedures and the design, layout and construction of the laboratory and its equipment. Audits generally use a detailed questionnaire to identify the organization's strengths and weaknesses. Following an audit, an action plan should be formulated and checks should be made to ensure that the plan is effectively implemented. There are a number of proprietary safety auditing systems available from safety organizations, consultants and some large chemical companies which could be used, or modified for use, in laboratories. The Health and Safety Executive's publication 'Monitoring Safety'[4] gives some examples of audit checklists and questionnaires which could be useful when devising a checklist for a laboratory.

14 Evaluation of Performance

The health, safety and environment performance should be evaluated at least annually and should be based on the information generated by the systematic approach advocated in this chapter. The evaluation should consider the standard of compliance with both the legal requirements and the laboratory's policies and procedures, the accident and ill-health record, and progress towards long term objectives. The results of the evaluation should be recorded, new objectives should be set and resources should be allocated so that the objectives can be met. A written evaluation is seen as a positive demonstration of the laboratory's approach and commitment to health, safety and environmental issues. The document can be used to show both senior management and employees the progress which has been made and how health, safety and environmental issues will be managed in the future.

15 Conclusion

The systematic approach to health, safety and environmental issues in the laboratory advocated in this chapter provides a framework to enable the Insured to show that he or she is a reasonable and prudent employer and has taken reasonable measures to provide good health, safety and environmental conditions which prevent the occurrence of accidents and occupational ill-health, and claims of negligence by employees.

Ten years ago at a previous RSC symposium on health and safety in the laboratory,[31] it was suggested that Insurers should reduce the cost of premiums for those employers with good health and safety records. Eagle Star are able to reflect an improved claims experience by offering the Insured a renewal bonus, on a sliding scale, of up to 25% of premium for a claim-free history.

16 References

1. Health and Safety Executive, Guidance Booklet HS(G)96 'The Costs of Accidents at Work', HMSO, London, 1993.

2. Health and Safety Executive, 'Effective Policies for Health and Safety: A Review Drawn from the Work and Experience of the Accident Prevention Advisory Unit of HM Factory Inspectorate', HMSO, London, 1980.

3. Health and Safety Executive, Occasional Paper Series: OP3, 'Managing Safety', HMSO, London, 1981.

4. Health and Safety Executive, Occasional Paper Series: OP9, 'Monitoring Safety', HMSO, London, 1985.

5. Health and Safety Executive, 'Success and Failure in Accident Prevention', HMSO, London, 1976.

6. British Standard 5750:1987, 'Quality Systems', British Standards Institution, London, 1987.

7. Health and Safety Executive, Guidance Booklet HS(G)65, 'Successful Health and Safety Management', HMSO, London, 1991.

8. Stokes v Guest, Keen and Nettlefold (Bolts and Nuts) Limited, (1968), 1, Weekly Law Report, 1776.

9. Management of Health and Safety at Work Regulations 1992, Statutory Instrument 1992, No. 2051, HMSO, London, 1992.

10. The Control of Substances Hazardous to Health Regulations 1988, Statutory Instrument 1988, No. 1657, HMSO, London, 1988.

11. 'Safe Practices in Chemical Laboratories', The Royal Society of Chemistry, London, 1989.

12. Manual Handling Operations Regulations 1992, Statutory Instrument 1992, No. 2793, HMSO, London, 1992.

13. Health and Safety (Display Screen Equipment) Regulations 1992, Statutory Instrument 1992, No. 2792, HMSO, London, 1992.

14. Personal Protective Equipment at Work Regulations 1992, Statutory Instrument 1992, No. 2966, HMSO, London, 1992.

15. Workplace (Health, Safety and Welfare) Regulations 1992, Statutory Instruments 1992, No. 3004, HMSO, London, 1992.

16. Provision and Use of Work Equipment Regulations 1992, Statutory Instrument 1992, No. 2932, HMSO, London, 1992.

17. 'Proposals for Fire Regulations (Places of Work) Regulations 1992 and Associated Guidance', Home Office, London, 1992.

18. Health and Safety Commission, Approved Code of Practice, 'Management of Health and Safety at Work: Management of Health and Safety at Work Regulations 1992', HMSO, London, 1992.

19. *Health and Safety Monitor*, 1992, **15**, (5), 1.

20. 'COSHH in Laboratories', The Royal Society of Chemistry, London, 1989.

21. 'Guidance on Laboratory Fume Cupboards', The Royal Society of Chemistry, London, 1990.

22. Health and Safety Commission, Approved Codes of Practice L5, 'Control of Substances Hazardous to Health and Control of Carcinogenic Substances: Control of Substances Hazardous to Health Regulations 1988', 2nd Edition, HMSO, London, 1990.

23. Health and Safety Executive, Guidance Note EH40/93 'Occupational Exposure Limits 1993', HMSO, London, 1993.

24. G.V. McHattie, M. Rackham, and E.L. Teasdale, *J. Soc. Occup. Med.*, 1988, **38**, 105.

25. R. Agius, *Ann. Occup. Hyg.*, 1989, **33**, 555.

26. E.V. Sargent and G.D. Kirk, *Am. Ind. Hyg. Assoc. J.*, 1988, **49**, 309.

27. British Standard 7750:1992, 'Environmental Management Systems', British Standards Institution, London, 1992.

28. R.W. Hazell, S.G. Luxon, and M.L. Richardson, in 'The COSHH Regulations: A Practical Guide', ed. D. Simpson and W.G. Simpson, The Royal Society of Chemistry, London, 1992, p. 145.

29. Accident Book, Form BI510, HMSO, London.

30. Reporting of Injuries, Diseases and Dangerous Occurrences Regulations 1985, Statutory Instrument 1985, No. 2023, HMSO, London, 1985.

31. 'Health and Safety in the Chemical Laboratory — Where do we go from here?', Royal Society of Chemistry, London, 1984.

The views expressed in this chapter are those of the author and do not necessarily reflect those of his employer.

CHAPTER 4

New and Forthcoming Legislation Associated with Laboratory Safety

H.P.A. ILLING

1 Introduction

This chapter identifies new health and safety legislation introduced in the early part of 1993, and also likely forthcoming legislation which could affect those working in laboratories. All this legislation and the associated Advisory Codes of Practice (ACOPs) and Guidance Notes (GNs) are the consequence of implementing European Community (EC) Directives.

Recent and future legislation may be grouped under four headings:

- The so-called 'six pack' (Table 1);

- The Regulations on the use or release of genetically modified organisms (Table 2);

- The introduction of the 'Carcinogens Directive' and 'Biological Agents Directive' requirements into the Control of Substances Hazardous to Health Regulations, ACOPs and Guidance;

- Proposed substance and product safety legislation.

2 The 'Six Pack'

This irreverent title has been given to a series of six sets of Regulations, two of which have associated ACOPs[1-2] and five of which have associated guidance.[3-6] The relevant Regulations are listed in Table 1, together with guidance information. They came into force on 1 January 1993. Although these Regulations cover all workplaces, clearly they have particular relevance to laboratories.

ACOPs and Guidance have different legal status. In the case of an ACOP, failure to comply with any provision of the Code is not in itself an offence, although that failure may be taken as meaning that a person has contravened the Regulations or sections of the Health and Safety at Work Act. However, it is open to that person to satisfy a court that the Regulations have been complied with in some other way. Other Guidance may give further information on what could be appropriate, but should not be regarded as an authoritative interpretation of the law.

Table 1 *The 'six pack'*

Title of Regulations	Title and type of supplementary information[a]	Refs.
Management of Health and Safety at Work Regulations 1992	Management of health and safety at work Approved Code of Practice	1
Workplace (Health, Safety and Welfare) Regulations 1992	Workplace health safety and welfare Approved Code of Practice and Guidance	2
Provision and Use of Work Equipment Regulations 1992	Work equipment Guidance	3
Personal Protective Equipment at Work Regulations 1992	Personal protective equipment at work Guidance	4
Manual Handling Operations Regulations 1992	Manual handling Guidance	5
Health and Safety (Display Screen Equipment) Regulations 1992	Display screen equipment at work Guidance	6

[a]The publication for each set of Regulations contains both the actual Regulations and the relevant guidance information.

3 Genetically Modified Organisms

Some laboratories either use genetically modified organisms as investigative tools in their work, or investigate and develop genetically modified organisms as part of their work. These laboratories may need to notify the appropriate regulatory authority under the Regulations[7-9] listed in Table 2, which came into force on 1 February 1993.

Table 2 *Regulations on genetically modified organisms*

Title	Coverage	Refs.
Genetically Modified Organisms (Contained Use) Regulations 1992	Health safety and environmental aspects of the contained use of genetically modified (by molecular biological techniques) micro-organisms (including cell cultures) and health and safety aspects of the contained use of genetically modified larger organisms (plants and animals).	7
Genetically Modified Organisms (Contained Use) Regulations 1993	Environmental aspects of the contained use of larger organisms.	8
Genetically Modified Organisms (Deliberate Release) Regulations 1992	Health safety and environmental aspects of the deliberate release and marketing of genetically modified (by molecular biological techniques) organisms.	9

These Regulations mirror Directives and do not therefore easily fall into single UK Governmental Department Responsibilities. Hence the 1992 'Contained Use' Regulations[7] were made under the Health and Safety at Work Act and the European Communities Act, and a small supplementary set of Regulations[8] under the Environment Protection Act was required to cover for possible environmental aspects of use of larger organisms not covered by the EC Directive. The 'Deliberate Release' Regulations[9] were set under the Environment Protection Act and the European Communities Act. Guidance on the 'Contained Use' Regulations and the associated notification and consent requirements has also been published.

4 Control of Substances Hazardous to Health Regulations

The Control of Substances Hazardous to Health (COSHH) Regulations were introduced in 1988. They were amended in 1992 in order to take into account the EC Directive on carcinogens at work, and to introduce a maximum exposure limit for bis(chloromethyl) ether. Included in the list of carcinogens are all the substances accorded the phrase R45 — 'May cause cancer' by the EC under labelling Directives.

This list is also incorporated in the draft 'Chemicals (Hazard Information and Packaging) Regulations' (see below) and its predecessors.

The two ACOPs 'Control of substances hazardous to health' and 'Control of carcinogenic substances' are bound into a single publication, the fourth edition of which covers these amendments. These ACOPs came into force on 1 January 1993.[10] It is also intended to implement the EC 'Biological Agents' Directive through an amendment to these Regulations and an Advisory Code of Practice, and the Health and Safety Commission has published a Consultative Document setting out its proposals.[11]

5 EC Directives on Substances and Products

The Health and Safety Commission has published 'Draft Proposals for the Chemicals (Hazard Information and Packaging) Regulations and associated Approved Documents.[12] As currently drafted, these so-called CHIP Regulations include ACOPs on the classification and labelling of substances and preparations dangerous for supply, and the provision of safety data sheets. These latter will contain specific information concerning the hazardous properties of a substance, appropriate handling information for different circumstances, and labelling requirements. The Approved Guide, which contains information on the classification system used to categorize hazards, also gives criteria by which the substances are categorized.

Some laboratories may make new chemicals. If they are placed on the market, notification may be required. The Health and Safety Commission has published draft proposals for a new set of Regulations to implement the EC 'Seventh Amendment Directive'.[13] These will update the Regulations for notification of 'New Substances' and set out a need for limited toxicity testing for all substances subjected to notification and an intention to undertake risk assessment for all full notifications.

6 Conclusion

Most of the recent and forthcoming legislation on health and safety in the workplace has emanated from the European Community as Directives. Laboratories are workplaces, thus this legislation is inevitably applicable to laboratories. The Directives are aimed at providing a common approach to health and safety issues throughout the European Community and, in many cases, the legislation is based on UK models. We can hope that, once a framework of health and safety legislation is implemented throughout the European Community, the pace of new introductions will be somewhat slower.

7 References

1. Health and Safety Commission, 'Management of health and safety at work. Approved Code of Practice', HMSO, London, 1992.

2. Health and Safety Commission, 'Workplace health, safety and welfare. Approved Code of Practice', HMSO, London, 1992.

3. Health and Safety Executive, 'Work equipment. Guidance on Regulations', HMSO, London, 1992.

4. Health and Safety Executive, 'Personal protective equipment at work. Guidance on Regulations', HMSO, London, 1992.

5. Health and Safety Executive, 'Manual handling. Guidance on Regulations', HMSO, London, 1992.

6. Health and Safety Executive, 'Display screen equipment work. Guidance on Regulations', HMSO, London, 1992.

7. Health and Safety Executive, 'A Guide to the Genetically Modified Organisms (Contained Use) Regulations. Guidance on Regulations', HMSO, London, 1993.

8. The Genetically Modified Organisms (Contained Use) Regulations 1993, Statutory Instrument 15:1993.

9. The Genetically Modified Organisms (Deliberate Release) Regulations 1992, Statutory Instrument 3280:1992.

10. Health and Safety Commission, 'Control of substances hazardous to health and Control of carcinogenic substances. Approved Codes of Practice', 4th Edition, HMSO, London, 1993.

11. Health and Safety Commission, 'The control of biological agents: proposals for amendments to the Control of Substances Hazardous to Health Regulations 1988', Health and Safety Executive, London, 1993.

12. Health and Safety Commission, 'Draft proposals for the Chemicals (Hazard Information and Packaging) Regulations and associated Approved Documents', Health and Safety Executive, London, 1992.

13. Health and Safety Commission, 'Draft proposals for the Notification of New Substances Regulations', Health and Safety Executive, London, 1993.

CHAPTER 5

The Classification of Carcinogens

P. WATTS

1 Introduction

According to a recent European Commission report,* one million people are afflicted with cancer each year, making cancer second only to heart disease as a cause of suffering. During the working years of life (20-64 years) cancer accounts for 40% of all deaths[1] and it is, perhaps, easy to understand why the general population is so apprehensive of chemicals classified as carcinogens (or for that matter mutagens or teratogens). Recent estimates suggest that cancer may cause the US Government $39 billion each year in lost production and income, medical expenses and research resources,[2] and the financial motive for reducing the cancer burden is, therefore, a powerful one.

Before this burden can be reduced, it is necessary to recognize the causes of cancer. Three environmental risk factors are implicated in the development of cancers:[3] exposure to *radiation, chemical* or *infectious agents*. Much experimental data for the involvement (and overlap) of these factors inevitably come from studies in animals. How can we identify and classify chemicals responsible for chemical carcinogenesis in man whilst maintaining a balanced perspective of the extrapolation of toxicological data to man and the benefits of chemistry in mitigating disease and malnutrition?

Ten years ago at another Royal Society of Chemistry symposium on laboratory safety[4] two recommendations (among thirteen) from the meeting expressed the reservations felt at the time about the interpretation and promulgation of toxicological data in the following terms:

'There is an immediate and continuing need for chemical societies to counter and discourage adverse media comments regarding health and safety in chemistry'.

'There is a longer-term need for an internationally accepted definition of a carcinogen to be established'.

---

*'Europe Against Cancer', a background report to an EC programme of action against cancer, 1987-1989, issued October 1st, 1987.

In this chapter we discuss the second of these concerns, and review and compare in some detail the various procedures for classifying carcinogens from toxicological data. We hope this exercise will benefit those who work with chemicals in a laboratory environment, and contribute to the debate on the role of chemicals in a modern society.

2 Occupational Cancer

Contemporary society faces the challenge of meeting the important need to reduce or prevent cancer. The administrative and scientific costs of such efforts require that the available resources be optimally focused in the most appropriate areas. These would seem to relate to the use of tobacco products, nutritional imbalances and other lifestyle factors.[5]

It is difficult to estimate the contribution occupational hazards make to total cancer risk, but the contribution is likely to be small. Sir Richard Doll has recently estimated that about 2% (maximum 6%) of cancer deaths are likely due to occupation.[1] This should be contrasted with estimates of 33-35% and 10-60% of cancer deaths attributable to tobacco and dietary factors respectively.[1] The gross occupational hazards observed in earlier years (nickel refining, manufacture of benzidine, 2-naphthylamine and asbestos textiles, bischloromethyl ether use, processes involving exposure to tar fumes and carcinogenic oils) and many of the lesser hazards (*e.g.* benzene and vinyl chloride exposure) have now been eliminated or controlled. Doll estimates that one liver angiosarcoma continues to occur each year as a result of past exposure to vinyl chloride, as well as some 50 or so cases of scrotal cancer which are probably due to coal tar, pitch or carcinogenic oil exposure, and 500-600 mesotheliomas from occupational asbestos exposure.[1] With the further passage of time since appropriate control measures were introduced, and the lack of evidence of major new hazards emerging despite intensive investigation, the proportion of cancer deaths due to occupational hazards is expected to fall. This expectation is likely to be realized provided that new chemicals are properly tested for potential carcinogenicity prior to widescale industrial use and provided that exposure to proven or suspected carcinogens is appropriately controlled.

3 Classification Schemes

One purpose of classifying carcinogens into categories is to provide a basis for making sound and sensible decisions on the use and handling of carcinogens. The criteria associated with a classification scheme should allow human, animal and other relevant data to be interpreted as to their relevance to humans in an appropriate and consistent manner.

Classification schemes for carcinogens based on qualitative criteria have been developed by a variety of expert national and international regulatory, industrial and independent groups, including the International Agency for Research on Cancer

(IARC), the European Community (EC) authorities, the US Environmental Protection Agency (EPA), and the American Conference of Governmental Industrial Hygienists (ACGIH). In addition, guidance on classification criteria has been developed by groups from the European Chemical Industry Ecology & Toxicology Centre (ECETOC) and BIBRA Toxicology International (BIBRA). A number of these classification schemes are summarized and discussed below. In addition to these qualitative schemes, some brief discussion is presented on existing efforts to rank carcinogens by potency of action, and to classify by broad consideration of mechanism.

Finally, we consider a Chemical Industries Association (CIA) scheme devised with the view to linking carcinogenicity ranking (based on qualitative and quantitative factors) with workplace hygiene controls.

4 The IARC Classification Scheme

Although the IARC classification scheme was not intended to serve as a direct basis for regulatory action, it is perhaps the most widely known scheme and the IARC monographs are recognized as an authoritative source of carcinogenicity information.

The IARC scheme evaluates qualitatively the *strength-of-evidence* for carcinogenicity from human and experimental animal data. In the first stage, a chemical is classified separately as to its carcinogenicity in humans and animals, the data being evaluated in each case as sufficient, limited, inadequate, or negative evidence. These categories refer only to the strength of the evidence, and not to potency or to mechanisms of action.

The next step is to collate other data relevant to an evaluation of carcinogenicity, such as information on preneoplastic lesions, tumour pathology, genetic effects, structural considerations, metabolism, pharmacokinetics and physicochemical parameters. Data on carcinogenic mechanisms are then evaluated, and the evidence (categorized as weak, moderate or strong) is assessed to see if an observed carcinogenic effect can be related to a particular mechanism. IARC then assesses the likelihood of that particular mechanism occurring in humans. (These mechanistic considerations are a recent modification to the IARC scheme.) Finally, an overall evaluation of the carcinogenicity to humans is made by reviewing the total body of evidence.

A chemical may be placed in **Group 1** (carcinogenic to humans), **Group 2A** or **2B** (probably or possibly carcinogenic to humans respectively), **Group 3** (not classifiable as to its carcinogenicity to humans), or **Group 4** (probably not carcinogenic to humans) by the criteria summarized in Table 1.[6]

Prior to the very recent modifications relating to mechanistic considerations (described below), the IARC scheme had traditionally relied solely on a *strength-of-evidence* approach. Such an approach begins with the presumption that, in the absence of adequate data on humans, it is biologically plausible and prudent to regard experimental animal carcinogens as presenting carcinogenic risk to man. The major contributions to strength-of-evidence evaluations are animal bioassays and epidemiology studies. More recently, genotoxicity data have also played a role in IARC evaluations. These approaches do not consider animal studies where there is no evidence of carcinogenic activity because these add nothing to the strength-of-evidence for carcinogenicity.

Table 1 *IARC classification scheme for carcinogenicity*

Group	Degree of evidence
1 Carcinogenicity to humans	a) Sufficient human b) Exceptionally — sufficient animal and strong evidence of a relevant mechanism
2A Probably carcinogenic to humans	a) Limited human and sufficient animal b) Inadequate human and sufficient animal, and strong evidence of a relevant mechanism c) Exceptionally — limited human only
2B Possibly carcinogenic to humans	a) Limited human and less than sufficient animal b) Inadequate human and sufficient animal c) Inadequate human and limited animal and other supporting relevant data
3 Not classifiable	a) Inadequate human and inadequate animal b) Inadequate human and limited animal c) Exceptionally — inadequate human and sufficient animal and strong evidence of an irrelevant mechanism d) Any not fitting into other groups
4 Probably not carcinogenic to humans	a) Evidence of lack of activity in humans and animals b) Inadequate human and evidence of lack of activity in animals and strong support by a range of other relevant data

The modifications allowing other data such as mechanistic relevance to be considered in the overall evaluation, means that IARC has shifted towards a *weight - of-evidence* approach. This approach weights indicators of lack of carcinogenicity along with indicators of carcinogenic activity, making a judgement concerning human hazard on a case-by-case basis rather than presuming human cancer risk from all substances showing carcinogenic activity in animals. The *weight-of-evidence* approach reflects a more complete evaluation of the total database than the *strength-of-evidence* procedure. This being the case, IARC's shift in approach from strength to weight-of-evidence is likely to be met with wide approval. A comparison between vinyl chloride and sodium saccharin has been cited as an example of the superiority of the weight-of-evidence approach. On a strength-of-evidence approach both are clearly carcinogenic in adequate animal studies. However, the weight-of-evidence approach suggests that they do not present the same degree of human hazard. (This is not a particularly good example because we have evidence of human carcinogenicity for vinyl chloride. Nevertheless, it is probably adequate for making the point.)

5 The EPA Classification Scheme

The US EPA guidelines on classifying carcinogens were originally published in 1986.[7] As with the IARC scheme, the EPA adopted a qualitative approach, but concentrated from the beginning on a weight-of-evidence approach rather than the strength-of-evidence approach originally adopted by IARC.

The EPA procedure involves an initial step of assessing the weight of evidence (though this stage is probably better described as a strength-of-evidence process) for carcinogenicity in humans and in animals, and classifying each as sufficient, limited, inadequate, no data, or no evidence (involving good negative studies). Other information (*e.g.* data from short-term tests or structure) can play a part in the classification, but potency does not. In step 2, human and animal classifications are combined to arrive at a tentative assignment to a category (see Table 2). The third and final step is to evaluate all relevant supportive data, including structure-activity relationships, short-term test findings, results of physiological, biochemical and toxicological observations, comparative metabolism and pharmacokinetic studies to determine whether the classification should be changed. This final step is where the weight-of-evidence approach takes over.

Moolenaar has expressed reservations over two features of the EPA scheme.[8] The first of these relates to step 2, where a tentative assignment is made before the complete data base has been evaluated. Thus, if carcinogenic activity has been seen in animals, a presumption of human risk may (perhaps unreasonably) be made at this step. However, this would not appear to prevent reclassifying into a more mechanistically sensible Group if the remaining relevant data, when evaluated, are sufficiently convincing as to support such a reclassification. The second issue Moolenaar raised is that the EPA has not yet described how to carry out the third and final step. The conclusion that positive animal studies lack predictability for humans appears to require epidemiology data showing non-carcinogenicity in man, the absence in humans of the biochemical pathway responsible for the animal tumours, and the unlikelihood of the mechanism behind the animal carcinogenicity occurring in man. Moolenaar points out that the first two criteria require proving a negative, a scientific impossibility. The third criterion, while theoretically possible to satisfy, is nearly impossible to meet in practical terms because of the uncertainty, at the molecular level, of the detailed mechanisms by which cancer occurs.

Table 2 *EPA classification scheme for carcinogenicity*

	Group	Degree of evidence
A	Human carcinogen	Sufficient human
B	Probable human carcinogen	
	1.	Limited human
	2.	Inadequate or no data or no evidence in humans and sufficient animal
C	Possible human carcinogen	Inadequate or no data or no evidence in humans and limited animal
D	Not classifiable	No or inadequate human and animal
E	Evidence of non-carcinogenicity	No evidence in at least two adequate animal tests, or in both adequate epidemiological and animal studies

6 The NTP Classification Scheme

The US National Toxicology Program (NTP) publishes an annual list of substances that are classified as either known or reasonably anticipated to be human carcinogens. Major sources of the data are the IARC monographs and the NCI/NTP carcinogenicity bioassay program. The available evidence is graded into categories such as sufficient, limited and inadequate. Known carcinogens are defined as substances for which the epidemiological evidence indicates a causal relationship between exposure and human cancer. Chemicals reasonably anticipated to be carcinogens are defined as having limited evidence from human studies and/or sufficient evidence from experimental animal studies.[2]

7 The EC Classification Scheme

7.1 EC Categories for Carcinogens

In the EC, the basic scheme for classifying carcinogens has been laid down in the framework of the EC regulations on classification and labelling of dangerous substances. Three categories are defined. Chemicals which do not require classifying can be considered as falling into a fourth (no classification) category. Category 1 substances are known human carcinogens. Category 2 substances are those with sufficient evidence to provide a strong presumption that human exposure may result in cancer, generally on the basis of appropriate long-term animal studies and/or other relevant information. These substances should be regarded as if they are carcinogenic to man. Category 3 substances are those that cause concern for man owing to possible carcinogenic effects but in respect of which the available information is inadequate. For these substances there is some evidence from appropriate animal studies, but this is insufficient to place the substance in Category 2.[9]

For a number of years, the EC gave no clear insights into the rationale they were using (and expected industry to use) for classifying carcinogens. In an attempt to provide some enlightenment, the European Chemical Industry Ecology & Toxicology Centre released in 1986 a technical report providing an industrial guide to the EC classification of carcinogens.[10] This group provided detailed criteria on how to evaluate and classify evidence from epidemiological studies, animal bioassays and other relevant test systems. Views on inclusion and exclusion criteria were given. It is not clear how the EC expert committees viewed the ECETOC document or to what extent this document has influenced EC thinking.

7.2 Proposals by BIBRA Toxicology International for Classifying Carcinogens

In an attempt to provoke further discussion on classification criteria and to provide objective assistance on the EC classification of carcinogens, BIBRA Toxicology International published in 1989 a proposed set of criteria in the Association's monthly BIBRA Bulletin which is circulated to all member companies. The criteria were intended to be applied only to the predefined EC classification and labelling scheme categories. As a scheme for carcinogen classification *per se*, an expanded scheme such as that developed by Ashby *et al.* 1990[11] (see later) would be more favoured. BIBRA's proposed classification criteria were as follows:

Category 1 — A positive relationship has been observed between exposure and human cancer in epidemiological studies in which chance, bias or confounding data could be ruled out with reasonable confidence. The existence of a causal relationship would be supported by any of:

a) An increased incidence of one or more cancer types in an exposed population in comparison with a non-exposed population;

b) Evidence of dose-time-response relationships, *i.e.* increased cancer incidence associated with higher exposure levels or with increasing exposure duration;

c) Association between exposure and increased risk observed in more than one study;

d) Demonstration of a decline in risk after reduction of exposure;

e) Specificity of any association, defined as increased occurrence of cancer at one target organ or of one morphological type.

Category 2 — Epidemiological data that are suggestive of a carcinogenic potential but that are not sufficient to satisfy the criteria for establishing causality. Reasons for inadequacy might include:

a) Uncertainty as to whether chance or bias may have contributed;

b) Inadequate control of confounding factors;

c) Only small numbers of cancer cases in the exposed population;

d) Unreliable information on levels or duration of exposure.

(Human data providing suspicions of carcinogenic potential may warrant a Category 2 tag irrespective of the nature of any animal data. Increased confidence in the credibility of a causal relationship would be provided by evidence of carcinogenicity in animals and/or genotoxic potential in short-term screening tests.)

— Cases where epidemiological data do not exist or appear negative but are actually inadequate for detecting carcinogenic potential (because of, for example, a small sample size, a short study duration or, as a result of excellent control practices, very low exposure levels)

and

evidence of carcinogenic potential in adequate animal studies carried out using exposure routes established as appropriate for detecting chemical carcinogenicity. Such evidence would normally comprise an increased incidence of malignant tumours (in some cases, it may be valid to combine benign and malignant tumours occurring at the same target tissue) in at least one mammalian species. If the carcinogen gave positive results in genotoxicity tests, it should be classified in Category 2 even if the doses used in the bioassay adversely affected the animals' physiological state, or produced recurrent injury in the tissue or organ in which tumours later appeared. For a genotoxin that gave a statistically and biologically significant increase in malignant tumours, classification in Category 2 is justified even if the background incidence in untreated controls is 'high' (a somewhat arbitrary distinction), even if (in general) the size of the increased incidence is small and even if tumours only appear late in the bioassay.

— Substances for which no data are available but which have very close structural links with established Category 2 carcinogens would often warrant the same classification.

Category 3 — There is a treatment-related increase in benign tumours only, without evidence of progression to malignancy.

— There is equivocal evidence of carcinogenicity from either limited or well conducted animal studies.

— The levels needed to induce tumours adversely affected the animals' physiological state, or produced recurrent injury in the tissue or organ in which tumours later appeared, and the chemical has shown no evidence of activity in an appropriate battery of short-term screening tests for genotoxicity.

— Substances for which no data are available but have very close structural links with established Category 3 carcinogens would often warrant the same classification.

Non-classification (that is, below the level of concern needed to have any labelling indication of carcinogenic potential).

— Strong support from good epidemiological studies (involving significant numbers of people, significant exposure levels and many years of exposure) of no carcinogenic potential

and

limited or no animal data.

— No evidence of carcinogenic potential in epidemiological studies that are limited in nature *and*

no evidence of carcinogenic potential in well-performed, comprehensive animal studies in all (at least two) species tested.

— Any evidence of animal carcinogenicity comes only from studies involving exposure by routes no longer considered appropriate for the detection of carcinogenicity

and

only local tumours occurred.

BIBRA supported the view that flexibility should be built into any system and that all chemicals should be examined on an *ad hoc* basis to achieve the most appropriate classification. It was believed important that the classification step on the path to regulatory compliance should be a relatively simple one to take since the responsibility for classifying and labelling substances would be so widespread throughout the industrial community.

7.3. EC Criteria for Classifying Carcinogens

More recently, official detailed criteria for classification within their three categories of carcinogens have been published by the EC.[12] Substances are placed into Category 1 on the basis of epidemiological data and into Categories 2 and 3 primarily on the results of animal experiments. For classification as a Category 2 carcinogen, either positive results in two animal species should be available, or clear positive evidence in one species together with supporting evidence such as genotoxicity data, metabolic or biochemical studies, induction of benign tumours, structural relationship with other known carcinogens, or data from epidemiological studies suggesting an association.

Two sub-categories make up Category 3. The first of these consists of substances that are well investigated but have given insufficient evidence of a tumour-inducing effect for classification in Category 2, and additional experiments would not be expected to yield further relevant information with respect to classification. The second sub-category comprises substances that are insufficiently investigated and have given inadequate data that raise concern for man. For these substances, classification is provisional on the basis that further experiments would provide more information allowing a final classification to be made.

The official guidance lists a number of relevant mechanistic arguments which help to distinguish between Categories 2 and 3 because they reduce the significance of experimental tumour induction for possible human exposure. The following are given as arguments that, especially in combination, would lead in most cases to classification in Category 3, even though tumours have been induced in animals:

- Carcinogenic effects only at very high dose levels exceeding the maximum tolerated dose.

- Appearance of tumours, especially at high dose levels, only in particular organs of certain species known to be susceptible to a high spontaneous tumour formation.

- Appearance of tumours, only at the application site, in very sensitive test systems (*e.g.* intraperitoneal or subcutaneous application of certain locally active compounds), if the particular target is not relevant to man.

- Lack of genotoxicity in short-term tests *in vivo* and *in vitro*.

- Existence of a secondary mechanism of action with the implication of a practical threshold above a certain dose level (*e.g.* hormonal effect on target organs or on mechanisms of physiological regulation, or chronic stimulation of cell proliferation).

- Existence of a species-specific mechanism of tumour formation (*e.g.* by specific metabolic pathways) which is irrelevant to man.

The guidance also provides a small number of mechanistic arguments relevant to distinguishing between Category 3 substances and those not requiring classification, on the basis that they exclude concern for man:

- A substance should not be classified in any of the categories if the mechanism of experimental tumour formation is clearly identified and with good evidence that this process cannot be extrapolated to man.

- If the only available tumour data are liver tumours in certain sensitive strains of mice, without any other supplementary evidence, the substance may not be classified in any of the categories.

- Particular attention should be paid to cases where the only available tumour data are the occurrence of neoplasms at sites and in strains where they are well known to occur spontaneously with a high incidence.

It is now clear that the EC classification system seems to utilize mechanistic data to a greater extent (or at least to spell them out in more detail) than do some other classifications.

8 The ACGIH Classification Scheme

The American Conference of Governmental Industrial Hygienists (ACGIH) is a professional organization whose members are primarily industrial hygienists working for local, state or federal government agencies. The Chemical Substances Threshold Limit Value Committee of the ACGIH, whose objective is to protect nearly all workers from harm by providing guidelines for occupational exposures, has identified two categories of carcinogen.

Group A1 carcinogens are confirmed human carcinogens, recognized to have carcinogenic or co-carcinogenic potential.

Group A2 carcinogens are suspected human carcinogens, based on either limited epidemiological evidence or demonstration of carcinogenicity in experimental animals. In classifying substances, the Committee considers epidemiologic studies, animal bioassays, short-term genotoxicity studies, studies of non-mammalian species and chemical structure analogy. Although not clearly a part of the classification process, the Committee recognizes the existence of potency variations, mechanistic differences, secondary metabolic pathways at extremely high doses and species-specific metabolism.[13]

9 The CEFIC* — CMA†/SOCMA‡ — CCPA§ Classification Scheme

This scheme was developed by a tripartite group of industrial experts from Europe, Canada and the USA.[14] The criteria for classification attempt to harmonize principles in the identification of carcinogens, provide guidance for the purpose of judging the relevance of experimental evidence using a weight-of-evidence approach, and serve as a constructive factor in discussions with authorities when differences in respective regulations are an issue.

The group has developed a classification system that is intended to be compatible with the major classification schemes currently employed, and is expanded to seven Categories:

Category 1   — Proven human carcinogens

Category 2   — Suspected human carcinogens
                    Based on suggestive epidemiology not sufficient to satisfy certain causality criteria **and** proven evidence from animal studies carried out under conditions that are relevant to human exposure.

---

*CEFIC, European Council of Chemical Manufacturers Federations;
†CMA, Chemical Manufacturers Association (USA);
‡SOCMA, Synthetic Organic Chemical Manufacturers Association;
§CCPA, Canadian Chemical Producers Association.

Category 3 — Proven animal carcinogens of potential relevance to humans.

Consideration should be given to this Category if the animal evidence is sufficient and was obtained under conditions relevant to the expected human exposure. Classification into this category would be supported by a clear statistical and biologically significant increase in the incidence of malignant tumours in an organ with a low spontaneous tumour incidence (or a large increase in incidence for tumours in organs having a high spontaneous tumour incidence), tumour induction at more than one site, a clear dose-response in the tumour incidence or time to tumour appearance, or tumour induction in more than one species or strain.

On the other hand, the data would not normally be considered sufficient for classification as a proven animal carcinogen if only benign tumours are induced, if the excesses are only of tumours such as liver nodules in rats and mice or pulmonary tumours in mice, if the excess is seen only in an organ having a high spontaneous incidence, if the route of administration is irrelevant, if the dose level required to induce tumours is so high that it adversely affects the normal animal physiology, or if no tumours are produced by exposure at or about those levels to which humans are likely to be exposed, even if tumours are found at exposures which produce chronic injury in the tissues or organ in which tumours later appear.

Information on metabolism, short-term test activity and (exceptionally) structure-activity relationships can be used to support decisions on the significance of data from other species and on classification.

Category 4 — Suspected animal carcinogens of potential relevance to humans.

For substances giving limited evidence from animal studies carried out under conditions that are possibly relevant to humans, without any other strong supporting evidence.

Experimental results that might lead to this categorization include a small increase in tumour incidence (*e.g.* where background data suggest that this could have occurred by chance, particularly if towards the end of the natural life span), an increase in only one species or strain (and which fails to meet certain defined biological and statistical criteria for Category 3 classification), an increase only in organs where the

natural incidence is high or variable, the use of irrelevant
exposure routes, non-relevance to humans (*e.g.* extensive
negative epidemiology), or an increase only of non-malignant
(which are known to progress to malignant) tumours.

Category 5   — Substances non-classifiable with regard to carcinogenicity.

For substances having some experimental evidence, but which
is limited in strength and/or is irrelevant to the human situation.

For example, where the only positive evidence is from animal
studies using excessively high doses that caused altered
physiological conditions such as tissue damage or a change in
metabolic pathway, and to which the positive result is
attributable; studies using exposure conditions that do not occur
when handled or used by humans; or studies using exposure
conditions in which the physical form tested was one to which
humans are not exposed. Other examples might include cases
where the experimental evidence is equivocal but the
epidemiological evidence is valid and negative; where the
positive studies were improperly designed or improperly
conducted, and could not be reproduced; where metabolism
(rates or products) is grossly different in the animals; and
where short-term data fail to support the animal data.

Category 6   — Negative evidence.

For agents that show no evidence of carcinogenicity in at least
two adequate (and appropriate) animal tests in different studies,
or in both epidemiological and animal studies.

Category 7   — No relevant data.

The scheme contains a number of sensible modifications to the more mature
schemes, and allows for additional flexibility. While a number of detailed
classification criteria are given, it is not always particularly clear how to interpret data
in terms of the phrase 'of potential relevance to man', and it is uncertain whether all
of the recommendations would find regulatory sympathy. For example, the group
suggest that 'for a substance to be considered a suspect carcinogen to humans or a
proven animal carcinogen of potential relevance to humans under expected conditions
of exposure, it will normally need to be genotoxic'. This view would seem to exclude
any non-genotoxic carcinogens of relatively high potency, where the threshold of
action may be below the human exposure level *e.g.* ethanol and certain substances
with hormonal (oestrogenic) properties. There is little guidance on how to evaluate

human information. Nevertheless, it is the initiative of groups such as this that provides the welcome impetus for improvements in the regulatory machine.

10 The Netherlands Classification Scheme

Prior to the publication of the EC classification scheme, the Netherlands developed a classification scheme that was rather different from others in use at the time.[15]

Category I — Complete carcinogens, initiators and substances that are seriously suspected of belonging to this category.

For substances giving positive results both in chronic bioassays and mutagenicity tests.

Category II — Substances acting by mechanisms such as promotion, hormonal alterations, non-specific stimulation, and suppression or over-stimulation of the immune system.

For substances giving positive results in chronic bioassays and negative results in mutagenicity tests. A non-genotoxic action has been proven.

Category III — Suspected carcinogens that cannot be classified according to biological mechanism.

For substances giving positive results in either mutagenicity tests or chronic bioassays.

Category IV — Carcinogenic substances that cannot be classified.

For substances giving positive results in chronic bioassays but there is no information on mechanism.

None — Not carcinogenic.

Negative results in bioassays of two animal species, and on absence of effects in mutagenicity studies.

This system is unusual in that classification relies initially on determination of the mechanism followed by consideration of the weight-of-evidence. One difficulty with this approach is that information on mechanism is not always available and, where available, is usually generated at a late stage in the investigation of a chemical's toxicological profile. It is not at all clear how to classify a substance into Category

III, and there seems to be no obvious spectrum in relative importance (in respect of hazard) with change in category. Where the Netherlands system has an advantage over other schemes is that the categorization does identify gaps in mechanistic understanding.

11 The AIHC Classification Scheme

This scheme was developed by the American Industrial Health Council in 1988 following the EPA's notice of intent to revise its carcinogen risk assessment guidelines. The basis for AIHC's comments was the view that a weight-of-evidence classification scheme was superior to one based on strength-of-evidence, and that science had moved beyond the presumption that all animal carcinogens represent a risk to humans.[16]

Category 1   — Known human carcinogen.

Category 2   — Animal carcinogen, probably carcinogenic to humans.

Evidence that experimental data are predictive for humans such as mechanistic or comparative metabolism data and/or confirmatory human evidence.

Category 3   — Animal carcinogen, possibly carcinogenic to humans.

Evidence such as positive bioassay data.

Category 4   — Animal carcinogen, limited evidence in animals.

For example: a single experiment; studies with unresolved questions of study design or non-physiological routes of administration; or results are of doubtful or questionable predictive value to humans such as agents that increase the incidence of benign or spontaneously occurring neoplasms.

Category 5   — Animal carcinogen, experimental evidence believed not to be of relevance to humans.

For example: rodent goitrogens; chemicals that induce only male rat kidney tumours; rat bladder calculi formers.

Category 6   — Unclassifiable.

Due to insufficient or no evidence.

Category 7 — Evidence of lack of carcinogenicity activity.

The AIHC classification procedure emphasizes the advantages of considering both the strength of conclusions from the animal bioassays and all of the data relevant to the appropriateness of the animal model as a predictor for humans in any particular case. There does appear to be unnecessary overlap between categories 3 and 4. In category 4, it is unclear why an increase in spontaneously occurring neoplasms should be seen as of 'doubtful or questionable predictive value', when most definitions of a carcinogen would seem to encompass such activity.

12 The Classification Scheme of Ashby *et al.*

Building upon the schemes of IARC, EPA, EC and the tripartite industrial group described earlier, a large group of expert industrial scientists has presented a classification scheme that takes account of new scientific knowledge about animal models in chemical carcinogenesis. Classification involves a weight-of-evidence approach to data from human, animal and mechanistic studies.[11]

Category 1 — Known human carcinogen.

Sufficient evidence for human carcinogenicity.

Category 2 — Carcinogenic activity in animals; probable human carcinogen.

Limited evidence of carcinogenicity in humans plus sufficient or limited evidence of carcinogenicity in animals together with data implying that the carcinogenic response in animals is relevant to a human response.

Inadequate evidence of carcinogenicity in humans and sufficient evidence of carcinogenicity in animals with strong additional evidence for relevance of the animal response to humans.

Category 3 — Possible human carcinogen.

Limited evidence of carcinogenicity in humans and limited evidence in animals or other supporting data.

Inadequate evidence of carcinogenicity in humans and sufficient evidence of carcinogenic activity in animals, but with inadequate information regarding relevance of the animal response to humans.

Strong evidence for lack of carcinogenic response in humans and sufficient evidence of carcinogenicity in animals together with data suggesting that the animal response is likely to be relevant to human response (this combination seems unlikely to occur very frequently).

Category 4   — Equivocal evidence of carcinogenic activity.

Inadequate evidence of human carcinogenicity and limited evidence of carcinogenic activity in animals with little or no information bearing on the relevance of the animal response to human response.

Category 5   — Evidence inadequate for classification.

Human data inadequate to draw a conclusion and animal data either inadequate or suggestive of non-carcinogenicity.

Category 6   — Carcinogenic activity in animals; probably not a human cancer hazard.

Human evidence bearing on carcinogenicity either inadequate or suggestive of non-carcinogenicity. Carcinogenic activity in animals either sufficient or limited, together with evidence from human or experimental studies that the carcinogenic response in animals is unlikely to be predictive of a human response.

Category 7   — Carcinogenic activity in animals; considered not a human cancer hazard.

Human evidence bearing on carcinogenicity either inadequate or suggestive of non-carcinogenicity. Sufficient or limited evidence for carcinogenic activity in animals, together with strong evidence from human or experimental studies that the carcinogenic response in animals is not predictive of a human cancer response.

Category 8   — Evidence of non-carcinogenicity.

Substances for which adequate, valid, information exists that indicates lack of carcinogenic activity. If any human evidence exists, it supports the conclusion that there is no association between exposure and an increased cancer risk.

The criteria for classifying within these categories are explained in detail and cover a comprehensive range of relevant information. The evidence for human and animal carcinogenicity is classified as sufficient, limited, inadequate or suggesting lack of carcinogenicity, and definitions of these descriptions are provided. The various types of corroborative information and the roles they play in the classification procedure are described in detail. From the bioassay, observations supportive of the predictivity of the animal cancer response include the same route of exposure, activity in several species, tumour site correspondence, multisite activity, activity in tissues analogous to human tissues, absence of cellular toxicity at target site, early tumour appearance, rapid tumour progression, activity at several exposure levels, lethality of tumours and rarity of the tumours at the target site.

Of observations other than the bioassay, supportive features include similar metabolic pathways/transformation, genotoxicity and DNA-reactivity of agent or metabolites, mechanisms known to occur in humans, structural similarity to a known human carcinogen, and an absence of disruption of homeostasis. The group provide a rationale for the role of each observation (and in each case a corresponding non-supportive observation is given) in supporting or detracting from the relevance to human hazard. Actual examples of the importance of individual observations are given in many cases.

In common with other schemes, this example recognizes that the animal models do not always reliably predict human responses. The scheme is comprehensive and flexible. Furthermore, it provides useful and helpful criteria that aid classification.

13 Non-Qualitative Classification Schemes

Though the emphasis of this chapter is on qualitative schemes of classifying carcinogens, we should touch briefly on other approaches. Specifically potency ranking and the division of carcinogens by genotoxic/non-genotoxic mechanisms are mentioned here.

13.1 Classification by Ranking Potency

One alternative to classifying carcinogens by strength or weight-of-evidence approaches is to classify on the basis of potency. Despite problems with this approach, potency estimations reflect the recognition that not all animal carcinogens pose equal threats to human health. To classify by potency, it is necessary to estimate the magnitude of carcinogenic activity in relation to dose.

13.1.1 Squire (1981)

One such scheme was proposed by Robert Squire, writing over 10 years ago from the Johns Hopkins University School of Medicine.[17] Squire employed six factors that should be used to produce an overall carcinogenicity potency score for chemicals that had been tested at multiple dose levels in at least two species. Substances merit 5 points for activity in one species or 15 points for activity in two or more species. One, two or three (or more) different neoplasm types are allocated 5, 10 or 15 points respectively. A further 1, 5, 10 or 15 points are awarded if the spontaneous incidence in control groups is > 20, 10-20, 1-10 or < 1% respectively. For activity at < 1 µg/kg bw/day, 1-1000 µg/kg bw/day, 1-1000 mg/kg bw/day or > 1 g/kg bw/day a further 15, 10, 5 or 1 point is allocated respectively. Finally, genotoxicity as measured in an appropriate test battery accounts for a further 25, 10 or 0 points for a description as positive, incompletely positive or negative respectively. Carcinogens are then ranked into one of five classes having scores of < 41, 41-55, 56-70, 71-85 and 86-100 total points. Aflatoxin and vinyl chloride (scores 100 and 90 respectively) fall into Class I while Class II would catch β-naphthylamine (score 81). Classes III and IV would include chloroform (score 65) and nitrolotriacetic acid (score 51), leaving Class V with carcinogens such as saccharin, DDT and chlordane (scores 36, 31 and 40 respectively).

On the face of it, Squire's scheme does not seem to deal with mechanistic data, other than genotoxicity. However, the literature suggests that epigenetic carcinogens in general demonstrate their activity in fewer species, at fewer target organs and at higher dose levels than genotoxic carcinogens. Thus in effect mechanism can influence the allocation of scoring points in more than one of the scoring categories.

As a confident understanding of a chemical's mechanism of carcinogenicity in laboratory animals and the relevance of this action to man is only achieved slowly and at great cost, Squire's proposed approach to potency may be sufficiently sophisticated to cope, for many years to come, with the vast majority of chemicals tested for carcinogenicity in at least two species.[18] Squire's approach puts the emphasis on test animal data in the belief that, without further knowledge of mechanisms, this information is the most relevant to risk.

13.1.2 ECETOC (1982)

An ECETOC Task Force has also evaluated the feasibility of ranking by carcinogenic potency,[19] and concluded that the provision of strict rules for potency categorization is not possible. However, some of the factors that could lead experts to judge that a carcinogenic potency was high were itemized, *e.g.* a large increase in the incidence of malignant tumours at low exposure levels; a large number of malignant tumours per animal; a short latency period; the development of malignant tumours after a small number of doses; tumour induction at a variety of organs; tumour induction in organs having a low natural incidence; the induction of a high incidence of malignant tumours in a number of strains and species; and ancillary information on mode of action, metabolism and tissue dose. The Task Force felt that potency estimates should be restricted to proven and putative human carcinogens and that, in the present state of knowledge, chemical carcinogens could be categorized only into broad categories of high, medium and low potency.

13.1.3 Extrapolation Models and Tumour Dose Calculations

Two other major methods for assessing potency merit mention here. The first involves the use of mathematical extrapolation models to predict carcinogenic risks at low doses; the second is the calculation of so called TD (tumour dose). Each uses a much more restricted range of data than do ECETOC or Squire — essentially just the dose-response data from a single animal carcinogenicity study.

The low dose extrapolation models use the carcinogenic response seen in laboratory animals exposed to relatively high doses and attempt to predict the carcinogenic risks at much lower dose levels within the likely human exposure range. At present, the various mathematical models have only a very general biological foundation, and it should perhaps be emphasized that values for risk generated by them are not experimentally verifiable.

Confidence in the model technology is further undermined by the fact that a very wide range of risk values in the low-dose regions of interest can be generated by applying different (but apparently equally rational, at least by our present understanding) mathematical formulae to the carcinogenicity data from a single animal study.

As an example of the wide range that can be obtained, it has been reported that using two different methods to process the data from one specific animal bioassay yielded estimated numbers of excess bladder cancer cases in a population of 200 million people ingesting 120 mg saccharin/day for 70 years ranging from 0.22 up to 1,144,000.[20]

Since the models produce such wide risk estimates, care should be taken over any risk management applications based on absolute risk values. Nevertheless, the models may have some role to play when attempting to rank the relevant risks posed by a number of carcinogens. In passing, it is worth noting the proposal that a modified one-hit procedure can be used to estimate the slope of the dose-response curve in the low-dose region, with the resulting figure being taken as an indication of potency.[21]

In the USA, the EPA considers that quantitative risk assessment is suitable for human carcinogens and probable human carcinogens. The model currently favoured by the EPA involves a linearized multistage procedure, which is based on the broad premise that cancer represents a number of discrete random biological events.

The other main scheme of potency evaluation uses only animal carcinogenicity data and is based on a TD concept. The TD_{50} is defined as the chronic dose rate (in mg/kg body weight/day) that would halve the actuarially adjusted percentage of tumour-free animals at the end of the 'standard' lifespan for the species. An approximate definition is the chronic dose that induces tumours in half of the tested animals over their lifetime. One group of scientists has published TD_{50} values on over 700 compounds from data on some 4000 long-term studies.[22-26]

TD values can be calculated for specific tumour types, for tumours at specific sites or groups of sites, or even for the overall tumour yield. A drawback of the TD_{50} approach is the reliance on only one datum. Two very different dose-response curves may generate the same TD_{50} value, and a single TD_{50} figure provides no information on the shape of the dose-response curve, and therefore of the potency, at the lower doses of interest.

In the UK, Department of Health scientists have concluded that the use of the TD_{50} as a ranking method is justified, and that analyses suggest there is a good correlation between carcinogenic potency in animal studies and potency in humans as assessed from epidemiology studies.[27] It was felt that the TD_{50} approach could be fine-tuned along the lines suggested in a working paper of the International Commission for Protection against Environmental Mutagens and Carcinogens (ICPEMC).[28] This fine-tuning included taking into account features such as the induction of tumours in tissues associated with high historical control incidence, tumour induction at multiple sites, tumour induction in both sexes of a species, and tumour induction in more than one species.

The ICPEMC group also developed the concept of the HADD (highest average daily dose that did not induce a statistical increase in tumours) as a measure of inactivity.[28]

14 Classification by Genotoxic or Non-genotoxic Activity

It is possible to ask whether it is feasible or useful to classify carcinogens using a simple genotoxic/non-genotoxic differentiation. Support for such a dichotomy relies on the consensus view that there may not be a level of exposure to a genotoxic carcinogen at which the risk of developing cancer is zero, whereas a non-genotoxic (epigenetic) mechanism of tumour induction may have a threshold, below which no risk exists. In theory then, there is an argument for considering genotoxic carcinogens to pose more of a cancer hazard than non-genotoxic carcinogens.

Unfortunately, this concept has been beset with difficulties. One such difficulty relates to the clear definition and understanding of what constitutes a genotoxin. There are agents that are directly DNA-reactive, and others that give positive results in short-term tests seemingly as a result of secondary effects on DNA-repair or synthesis mechanisms. In some cases it seems that the DNA is itself altered but not at a particularly early stage in a proposed mechanism to tumours. In recent times more emphasis has been placed on the performance of *in vivo* short-term tests both for defining a genotoxin and as a more reliable predictor of carcinogenic activity. Few genotoxins give positive results in all short-term test systems so a judgement of 'genotoxicity' must depend on the availability of a battery of test data and on a weight-of-evidence approach. Some genotoxins may have no demonstrable carcinogenic activity (*e.g.* due to metabolic deactivation) and it may be that some such genotoxins could induce tumours by epigenetic mechanisms.

All in all, a classification scheme based solely on a distinction between genotoxic and non-genotoxic potential is unlikely to be as straightforward to devise as it may first appear, and would perhaps not be particularly useful on its own. Information on mechanism would seem to be better used as a component of a weight-of-evidence process employed within such qualitative classification and potency ranking schemes as have been discussed in this chapter.

15 Classification of Carcinogens — Implications for Handling

Careful planning, implementation and monitoring of good industrial hygiene practices, and sound engineering controls can provide effective limitation of occupational exposure to carcinogens. The components of a carcinogen control progress include exposure monitoring and surveillance, medical surveillance, regulated and controlled access areas, protective clothing and equipment, housekeeping, hygiene facilities and practices, employee information and training, signs and labels, process enclosure, process modification, process automation and ventilation.[29]

The success of a hygiene control programme depends largely on simple, proven industrial hygiene concepts. Certain of these practices (*e.g.* personal hygiene, good housekeeping and good work practices) are particularly important in the control of carcinogens, though the most effective means of limiting exposure relies on an engineering control (an automated closed system). Controlled access and provision of adequate personal hygiene facilities are also essential in minimizing worker exposure.

The engineering controls, work practices and personal protection factors actually selected will depend on the specific carcinogen to be controlled and its physical, chemical and toxicological properties. Identification and classification of carcinogenic hazard is only the first step in defining safe handling practices. It seems reasonable to believe, however, that not all carcinogens pose equal carcinogenic risks. Firstly, there is reason to think that known human carcinogens vary in potency. Second, what we know about the mechanisms of tumour induction and genotoxicity profiles leads us to believe that not all animal carcinogens will pose similar degrees of risk to humans exposed generally at much lower levels. With safety evaluation studies uncovering a larger proportion of animal carcinogens than had been anticipated, some means of prioritizing and focusing resources on the appropriate (higher risk) carcinogens would seem to be essential. The known wide variations in potency estimations (*e.g.* some eight orders of magnitude may separate saccharin from a nitrosamine) provide some support for the introduction of ranking schemes within the categories of carcinogens.

16 The CIA Ranking and Hygiene Control Scheme

Recently a Working Group of the Chemical Industries Association proposed a scheme for ranking the carcinogenicity of aromatic amines and nitro compounds.[30,31] The scheme includes both qualitative (weight-of-evidence) and quantitative (carcinogenic potency based on TD) factors, and was drawn up specifically with a view to linking up with hygiene controls in the workplace.

In the first stage, a chemical is classified in the fairly traditional qualitative manner as a proven human, suspect human, proven animal or suspect animal carcinogen, as non-classifiable or as a chemical for which there is 'negative evidence' from good quality studies:

Proven human   —   This categorization would result from the conclusions of an IARC evaluation if consistent with other known, valid data, or from good quality epidemiological studies showing a cause and effect relationship.

Suspect human   —   Comprising proven animal carcinogens that have given suggestive epidemiological evidence, *e.g.* case reports, clusters. This group might also include chemicals giving suggestive epidemiology with only limited positive animal studies, if reinforced by positive *in vivo* genotoxicity data.

Proven animal   —   Comprising chemicals that induce statistically significant increases in malignant tumours, which should be apparent in more than one species. Where data exist for one species only, careful consideration of tumour type would be needed before placement in this group, and genotoxicity results would be influential. Support for inclusion in this group would be provided by positive *in vivo* genotoxicity and evidence of a dose-response relationship (presumably in the bioassay).

Suspect animal   —   Indicated by conflicting and/or limited evidence, or positive studies of good quality but with results of questionable relevance, *e.g.* tumours only induced at extremely high doses (> 1 g/kg bw/day) or at doses that disrupt the normal functioning of the target organs.

Non-classifiable   —   Indicated by animal data of doubtful significance (*e.g.* injection or implantation site tumours) or effects known to be due to secondary mechanisms that are clearly understood. Chemicals may also be allocated to this group where there are doubts over purity of the tested material or evidence that an atypical material was tested.

Negative evidence   —   Indicated by negative results in high quality studies.

In the second phase of the exercise, a quantitative measure of potency (in this case the TD_{50}) is obtained and applied. In selecting a specific TD_{50} value from those available, the group recommends looking for clear increases in the incidence of malignant tumours (though combinations of benign and malignant tumours may be used if they arise from the same cell type), with most weighting being given to the malignant tumour type showing the highest numerical increase compared with the control group. In their validation exercise, the authors utilized the lowest TD_{50} (reflecting the most potent carcinogenic effect).

Potency categories are then defined, with cut-off ranges at < 1, 1-10, or 10-100 mg/kg bw/day for suspect animal carcinogens, and < 10, 10-100 or 100-1000 mg/kg bw/day for proven animal carcinogens, while suspect human carcinogens are divided into two groups on the basis of whether the TD_{50} in the animal study was less or more than 100 mg/kg bw/day. The CIA scheme then provides a matrix that allows a carcinogenic chemical's weight-of-evidence and potency to determine its place in one of five carcinogenicity (and two other) classifications (Table 3).

Table 3 *Matrix for ranking carcinogenicity of aromatic amines and nitro compounds*[a]

	Weight of evidence	*TD_{50} potency category*		*Classification*
1	Proven human	Any		Proven human
2	Suspect human	<100 mg/kg/day	}	
	Proven animal	<10 mg/kg/day	}	A
	Suspect animal	<1 mg/kg/day	}	
3	Suspect human	>100 mg/kg/day	}	
	Proven animal	10- 100 mg/kg/day	}	B
	Suspect animal	1-10 mg/kg/day	}	
4	Proven animal	100-1000 mg/kg/day	}	C
	Suspect animal	10-100 mg/kg/day	}	
5	Suspect animal	100-1000 mg/kg/day		D
6	Non-classifiable			*Incapable of classification
7	Negative evidence			

*Refer to expert
[a]Taken from Crabtree *et al.* 1991, reference 31.

To achieve the major objective of devising categories of safe handling guidelines, the Working Party's next step was to allocate these rankings to one of four occupational hygiene control strategies (Table 4).

Table 4 *Hygiene control strategies based on carcinogenicity ranking*[a]

Carcinogenicity Ranking Class	Hygiene Control Strategy
Proven human and A	Level 4
B	Level 3
C and D	Level 2
Negative evidence	Level 1 (basic occupational hygiene strategy)

[a]Adapted from CIA, 1992, reference 30.

Details of these control strategies are provided in CIA, 1992 and Crabtree *et al.*, 1991 (references 30 and 31). The ranking procedure allows room for flexibility and for toxicological expertise to influence decision making within the system, and some worked examples demonstrate how discussions of dose-response, excessive mortality, historical control data and biological significance can have an impact in the qualitative assessment. The authors emphasize that the scheme deals only with carcinogenic potential, and that other toxic effects may dictate the final hygiene control strategy.

Chemicals should be considered to have the potential for multispecies carcinogenicity if there are good quality positive data from well-validated *in vivo* genotoxicity studies (such as the mouse micronucleus test and the liver unscheduled DNA synthesis assay) carried out to acceptable current standards and using a physiologically normal route.[31] The scheme is intended for use by 'non-toxicologists having some familiarity with toxicology', though experts may have to be consulted on the more sophisticated issues, such as deciding which tumour types and incidence patterns are critical for qualitative classification. Though the focus on this occasion was aromatic amines and nitro compounds, the devised scheme would seem to be equally useful for other chemical candidates.

Indeed, more recently Money from ICI has described a structured approach to occupational hygiene in the design and operation of fine chemical plant, using the Crabtree *et al.* ranking scheme.[32]

17 The Laboratory Chemist

In general, the risks of developing cancer from exposure to a carcinogen increase with increasing duration of exposure. Since industrial workforces may be handling a particular carcinogen for many years, it is probable that, all other things being equal, the laboratory chemist may be at a lower risk of developing cancer from the occasional handling of a carcinogen. A recent study of the mortality of professional chemists in England and Wales for the period 1965-1989[33] appears to support this view. There was an overall low mortality rate with fewer deaths than expected from cancers.[33] However, there was an excess mortality from certain types of cancer, notably lymphatic and haematopoietic cancers.[33]*

Workers in chemical manufacturing factories have been the focus of most of the efforts to evaluate and manage carcinogenic risks. While good industrial practices are encouraged in these workers, engineering controls are a major element in controlling and restricting exposures. The emphasis on engineering controls and employer practices is encouraging, since it helps to minimize the risk that those workers who are careless in handling chemicals could lower occupational health standards.

The research laboratory chemist is in a rather different position. He or she may be one of only a very small group (perhaps the only one) of employees handling a particular carcinogenic chemical. The carcinogen may be handled on only a very few occasions. The level of engineering control available may be much lower than that used in an industrial process. Finally, the research laboratory chemist might be expected to have a better understanding of the physicochemical and toxicological properties of a chemical than the average industrial process worker. Almost certainly he or she will be supervised to a lower degree.

Inevitably then, the lone chemist or those in a small research group must take a higher degree of individual responsibility for their own safe working practices and, ultimately, the level of risk their activities pose to their own health and the health of those who work nearby. Knowledge of a chemical's intrinsic carcinogenic properties provides the power the chemist needs as the first phase in ensuring that health is not compromised at work. Classification schemes based on qualitative factors, including weight of evidence, and potency can be useful tools to the research chemist selecting or devising a protocol for handling a chemical carcinogen in the laboratory.

18 Conclusions

Over the years we have seen how carcinogen classification schemes have increased in scientific sophistication, one consequence being a move away from a strength-of-evidence approach towards one based on weight-of-evidence. Our understanding of mechanisms, genotoxicity and species-specificity has been influential in modifying the approach. Improved industrial control will also mean that the strength-of-evidence approach will become increasingly less useful, as chemicals will be given fewer opportunities to demonstrate any potential carcinogenicity in humans. A recognition

---

*Refer to Chapter 1 for more details of this study, carried out by The Royal Society of Chemistry. An epidemiological survey of cancer risk in biology research laboratory workers in nine European countries is currently in progress, and the results should be available in 1995. (IARC Biennial Report, 1990-1991).

of the importance of potency considerations is also encouraging; we must acknowledge that not all carcinogens pose the same level of risk under similar exposure conditions. In the meantime, it would be useful to improve harmonization between the various classification schemes. For example, Moolenaar has noted that some chemicals are classified as probable human carcinogens by the EPA, as possible human carcinogens by IARC, but not even classified as carcinogens by the EC.[8]

Titles of categories should be accurate but should also provide adequate communication. For example, the title 'probable' may imply 'outcome' in the mind of the general public but 'potential' in the mind of the evaluating scientist.

To re-emphasize the points made by many others, proven or suspected carcinogens should not be used if suitable substitutes exist. When carcinogens have to be used, exposure should be reduced to the minimum. It should be borne in mind that occupational hazards are believed to account for only a small proportion of the total numbers of cancers. With good training, data provision, careful handling practices, good occupational hygiene, adequate research funds, proper engineering controls and appropriate monitoring, supervision and medical surveillance, we should be optimistic that the numbers of cancers arising from occupational exposures will be reduced even further.

19   References

1. R. Doll, in 'Cancer in the Workplace. Reducing Risk and Promoting Health', Conference Proceedings, 15 October 1992 at SCI Headquarters, Health and Safety Executive, Europe against Cancer, Society of Chemical Industry, London, 1992, pp. 8-11.

2. Sixth Annual Report on Carcinogens, 1991 Summary, DHHS Public Health Service, National Institute of Environmental Health Sciences, 1992.

3. J.A. Wyke, in 'Introduction to the Cellular and Molecular Biology of Cancer', ed. L.M. Franks and N. Teich, Oxford University Press, Oxford, 1986, p. 176.

4. 'Health and Safety in the Chemical Laboratory — Where do we go from here?', The Royal Society of Chemistry, London, 1984, pp. 197-198.

5. M.C. Pike, in reference 3, p. 72.

6. 'Solar and Ultraviolet Radiation', IARC Monographs on the Evaluation of Carcinogenic Risks to Humans, Volume 55, Lyon, 1992.

7. US Environmental Protection Agency, *Federal Register*, **51**, 33992, 1986.

8. R.J. Moolenaar, *The SIRC Review*, 1990, **1**, 43.

9. Commission Directive 83/467/EEC of 29 July 1983. *Official Journal of the European Economic Communities,* **26** (L257), 1, 16 September, 1983.

10. 'Technical Report No. 21. A Guide to the Classification of Carcinogens, Mutagens & Teratogens under the Sixth Amendment', European Chemical Industry Ecology and Toxicology Centre, Brussels, 1986.

11. J. Ashby, N.G. Doerrer, F.G. Flamm, and 13 others, *Regulatory Toxicology and Pharmacology,* 1990, **12**, 270.

12. Commission Directive 91/325/EEC of 1 March 1991. *Official Journal of the European Economic Communities,* **34** (L180), 1, 8 July, 1991.

13. R. Spirtas, M. Steinberg, R.C. Wands, and E.K. Weisburger, *Am. J. of Public Health,* 1986, **76**, 1232.

14. E.J. Sowinski, B. Broecker, J. Faccini, and 14 others, *Regulatory Toxicology and Pharmacology,* 1987, **7**, 1.

15. 'Report on the Evaluation of the Carcinogenicity of Chemical Substances', Health Council of the Netherlands (HCN), Govt. Publishing House, The Hague, 1980.

16. 'Initial Comments by AIHC concerning EPA's Notice of Intent to Revise its Carcinogen Risk Assessment Guidelines', American Industrial Health Council, Washington, October 11, 1988.

17. R.A. Squire, *Science, N.Y.,* 1981, **214**, 877.

18. J. Hopkins, *Food. Chem. Toxicol.,* 1989, **27**, 417.

19. 'Monograph No. 3. Risk Assessment of Occupational Chemical Carcinogens', European Chemical Industry Ecology and Toxicology Centre, Brussels, 1982.

20. I.C. Munro and D.R. Krewski, *Food. Chem. Toxicol.,* 1981, **19**, 549.

21. E. Crouch and R. Wilson, *J. Toxicol. Environ. Health,* 1979, **5**, 1095.

22. L.S. Gold, C.B. Sawyer, R. Magaw, and 9 others, *Environ. Health Perspec.,* 1984, **58**, 9.

23. L.S. Gold, M. de Veciana, G.M. Backman, and 7 others, *Environ. Health Perspec.,* 1986, **67**, 161.

24. L.S. Gold, T.H. Slone, G.M. Backman, and 5 others, *Environ. Health Perspec.,* 1987, **74**, 237.

25. L.S. Gold, T.H. Slone, G.M. Backman, and 6 others, *Environ. Health Perspec.*, 1990, **84**, 215.

26. L.S. Gold, T.H. Slone, N.B. Manley, and 4 others, *Environ. Health Perspec.*, 1991, **96**, 11.

27. K.N. Woodward, A. McDonald, and S. Joshi, *Carcinogenesis*, 1991, **12**, 1061.

28. S. Nesnow, *Mutat. Res.*, 1990, **239**, 83.

29. J. Singh, *Appl. Ind. Hyg.*, 1988, **3**, 58.

30. 'Safe Handling of Potentially Carcinogenic Aromatic Amines and Nitro Compounds', Chemical Industries Association, London, 1992.

31. H.C. Crabtree, D. Hart, M.C. Thomas, B.H. Witham, I.G. McKenzie, and C.P. Smith, *Mutat. Res.*, 1991, **264**, 155.

32. C.D. Money, *Annal. Occup. Hyg.*, 1992, **36**, 601.

33. W.J. Hunter, B.A. Henman, D.M. Bartlett, and I.P. Le Geyt, *Am. J. Ind. Med.*, 1993, **23**. 615.

CHAPTER 6

Methods for the Disposal of Carcinogens and Carcinogenic Waste

M. CASTEGNARO

1 Introduction

The international effort devoted to cancer research requires experimental work with carcinogenic or mutagenic substances in an increasing number of chemical, biochemical, and biological laboratories. These investigations may result in accidental spillages, and will produce hazardous waste which will need to be decontaminated before being released from the laboratory environment.

Treatment of cancer and other diseases also involves handling substances which have been classified as human carcinogens,* or probable/possible human carcinogens.†[1,2] This therapeutic work similarly generates toxic waste and spillages which require decontamination. In addition, excreta and biological fluids from patients administered cytostatic drugs should be classified as hazardous, since not only the original drug but biologically active metabolites may be present. Equipment which has been used to administer carcinogenic drugs or has been in contact with biological fluids must also be decontaminated as it could otherwise present a hazard, to indirectly exposed personnel.[4]

When initiating work with carcinogenic substances, it is important to set up a plan for dealing with waste. Such a strategy must include emergency plans for treatment of spills, and a written procedure for degrading each class of carcinogen to be handled.

'Good laboratory waste management begins with preventive measures, that is, identification of steps that can be taken to reduce the volume of chemicals that enter the waste disposal process and to prevent unusual, difficult disposal problems' is how Joyce[5] summarizes the first step to be taken when handling any laboratory chemical, and especially noxious substances such as chemical carcinogens. This approach is also stressed in the US publication 'Prudent Practices for Disposal of Chemicals from Laboratories',[6] which recommends that procedures should be in place for the accumulation and disposal of wastes before the wastes are generated.

---

*Cyclophosphamide, melphalan, semustine, thiotepa, azathioprine, treosulphan, chlornaphazine, chlorambucil, cyclosporin, analgesic mixtures containing phenacetin.
†Adriamycin, azacytidine, chloramphenicol, BCNU, CCNU, chlorozotocin, cisplatin, phenacetin, procarbazine, bleomycin, dacarbazine, daunomycin, merphalan, methylthiouracil, metronidazole, mitomycin C, nafenopin, niridazole, panfuran S, phenobarbital, progestins, streptozotocin, trichlormethine.

Accumulated waste should be properly identified and separated into different categories. Liquid waste must be segregated from solid waste. Each container of liquid waste should be labelled with the following information: identity of laboratory or person producing the waste; date; type of carcinogen; approximate concentration; and solvent (aqueous or organic, inflammable or not, specified if possible). Solid waste should also be properly labelled. It is essential that the label on any waste contains sufficient information to assure safe handling and disposal.

The waste producer has a moral and legal obligation to ensure that the waste is handled and disposed of in ways that pose a minimum potential hazard to health and the environment, both in the short and long term. This became a legal obligation in the USA in 1980, when the US Environmental Protection Agency put into effect federal regulations on hazardous waste management.[6] In France, several laws have been enacted[7-8] concerning responsibility for the transportation and disposal of wastes.

Accumulation and storage of waste can only be a temporary measure and the decision on the choice of appropriate disposal methods should not be delayed.

2 Strategies for the Disposal of Carcinogenic Waste

Though a number of books give general guidelines and procedures for the disposal of chemical wastes and for handling spillages,[9-14] only recent publications describe methods for handling carcinogenic substances.[6,15-18]

In this chapter aspects of waste disposal pertaining to chemical carcinogens are reviewed, and the results of a programme on the degradation of carcinogens initiated by the International Agency for Research on Cancer are summarized in an Appendix. This programme was jointly supported by the Office of Safety of the National Institutes of Health in the USA in conjunction with NCI Frederick Cancer Research Facility, the French Ministry of the Environment, and 51 scientists from 13 countries.

2.1 Evaporation and Disposal through the Sewage System

Whilst it may be possible to dispose of chemical carcinogens by evaporation into the atmosphere or through the sewage system (techniques which were used in some research laboratories up to about 15 years ago), it is inadmissible to proceed in this way *unless* there is strong evidence that the compounds are extremely rapidly and completely degraded into non-toxic or non-genotoxic substances.

Hospitals may have, however, little or no choice but to use the public sewage system for the disposal of excreta of patients treated with chemotherapeutic agents. The excreta may contain not only the original drug but also active metabolites.

2.2 Burial of Chemical Carcinogens

Recommendations have been put forward by the US Environmental Protection Agency[19] for the selection of an appropriate site. Specific factors to be evaluated include:

i)   The chemical characteristics and amounts of hazardous wastes;

ii) The potential for release into the environment;

iii) The sensitivity of the particular environment to the hazardous wastes;

iv) The proximity of the hazardous wastes to humans;

v)   The potential effects of wastes on human health.

Methods have been proposed to assess the health effects at chemical disposal sites from the surveillance of various geohydrological parameters.[20]

For cytostatic wastes, the US Occupational Safety and Health Administration (OHSA) propose that 'cytostatic wastes must be handled separately from other hospital trash and must be regarded as toxic (hazardous) wastes and disposed of in accordance with applicable regulations. It is recommended that disposal should take place in licensed sanitary landfill using the services of a fully licensed firm.'[21]

Several reports have demonstrated the dangers of burying chemicals, for example, the infamous case of Love Canal.[22] It should also be noted that not only the buried substances can cause problems, but also that their interactions can generate noxious substances, as demonstrated in Japan,[23] where the presence in a landfill site of phosphine, a colourless poisonous gas,[24-25] was attributed to the reaction amongst phosphorus-containing carbides, potash and water. Mutagenic activity has also been demonstrated in leakage of municipal solid waste landfill.[26]

As a result of the increasing difficulty in finding suitable landfill sites, some waste disposal contractors take advantage of the economic difficulties and lack of environmental legislation in developing countries to export their waste.[27-31] Such traffic must be so severely penalized that it becomes dissuasive for contractors to by-pass the rules. The Basel Convention[32] attempts to regulate this aspect of waste management.

2.3 Incineration of Chemical Carcinogens

Incineration is a method with some potential for the efficient and safe disposal of organic chemical waste. The performance of incinerators, however, can vary. For similar equipment, the efficiency of the incineration process may differ due to a number of parameters, such as retention time of the compounds, feed rate, and air supply to furnaces. Most of the studies on the performance and risks from incineration have been carried out on municipal solid waste incinerators. Several analyses on the risk of discharges have been published,[33-35] but details of the conditions of incineration are usually omitted. Numerous carcinogenic contaminants have been found in the gaseous and particulate emissions of municipal refuse incinerators, including polychlorinated dibenzo-p-dioxins (PCDDs), polychlorinated dibenzofurans (PCDFs), polycyclic aromatic hydrocarbons and some heavy metals.

By selecting the operating conditions, these emissions can be minimized. For example, temperature is an important factor in dioxin formation and destruction in an incinerator; the formation temperature is around 500 °C and the destruction temperature is at least 900 °C in a well oxygenated furnace with a residence time of the toxic gases of at least 1 sec.[36] With metal emissions, temperature is not the only factor. The physico-chemical properties of the elements will influence their partitioning between flue gas, dust and slag. In an incinerator running at 900 °C, 72% of mercury is in the flue gas, but only 12% of cadmium, 5% of lead, and 1% of copper. Another important factor which will affect the level of elements in the flue gas is the concentration of halogens, and in particular chlorine. This can influence the emission of volatile metal chlorides, such as cadmium, zinc or mercury chlorides.[37]

From these data on incineration of municipal waste, it is apparent that low temperature incineration should not be used for laboratory waste and especially carcinogenic waste. Only a few studies have been reported on the efficiency of degradation of chemical carcinogens in incinerators. Some work conducted at the National Center for Toxicological Research, Food and Drug Administration (Jefferson, USA), has shown that 2-acetylaminofluorene can be destroyed by incineration at 737 °C, measured in the secondary chamber of the incinerator, with a minimum residence time of 2 s (cited by Barbeito).[38]

In an investigation of incineration conditions, the thermal decomposition of some polycyclic aromatic hydrocarbons and some halogenated analogues was studied.[39] Even at temperatures as high as 725 °C, 10-19% of naphthalene compounds were still present, and 0.21% of benzo[e]pyrene. In municipal waste incinerators, PCDDs and PCDFs are stable at temperatures up to 800 °C, and at temperatures of 300-500 °C, these compounds can be formed from chlorinated precursors.[40]

Wilkinson and Rogers[41] evaluated five incinerators for their capacity to degrade a standard charge of 18 non-carcinogenic compounds by monitoring the exhaust for both original charge and NO_x, SO_x, C_xH_x, HCl, CO, and CO_2. Three of the units gave satisfactory results, but the authors concluded that further research was required before recommending the use of these incinerators for the destruction of carcinogens, mutagens, and teratogens. A number of such studies would be necessary, not only to assess the efficiency of incineration to degrade chemical carcinogens, but also to determine whether the residues produced, either in the gas stream or in aqueous effluents, are non-carcinogenic. Chien and Thomas[42] have suggested that volatile molecules such as nitrosamines may not be completely degraded by incineration.

The efficiency of incinerators might be improved by equipping the chimney with low temperature catalytic systems, which have been shown to degrade efficiently some polycyclic aromatic hydrocarbons and aflatoxins.[43]

Particulate emissions can be reduced by using either an electrostatic precipitation system or a water scrubbing system. The former method produces dry particulates which may be contaminated, whilst the latter gives contaminated sludge and water. Combination of both techniques should improve the quality of exhaust air, and the waste water can then be neutralized.

In conclusion, high temperature incinerators with proper control of the feed rate, air flow, residence time, and scrubbing systems of the flue gas can be regarded as a useful method for the disposal of laboratory wastes contaminated with chemical carcinogens.

In view of the paucity of information available in the literature on the incineration of carcinogenic wastes plus the data from high temperature municipal waste incinerators, we decided, at IARC, to adopt a temperature of 1100 °C in an incinerator equipped with a burner, a post-combustion burner, and a water scrubbing system.

2.4 Chemical Destruction of Chemical Carcinogens

It is possible to alter or even destroy hazardous substances by chemical modification. Methods have been suggested for treating, for example, acids, bases, mercaptans, acid halides, anhydrides, aldehydes, ketones, and amines.[17] Until the late 1970s little work had been reported on chemical methods to decontaminate carcinogenic substances.[44] The IARC group which evaluated the data available then recommended that:[44]

i) 'Studies are needed to evaluate the efficiency of methods for decontaminating equipment, including liquid traps, filters, glassware, and animal cages';

ii) 'Research into methods for the destruction and disposal of chemical carcinogens is urgently needed'.

This prompted IARC and the US Office of Safety of the National Institutes of Health to join forces, and set up a programme for the evaluation of chemical methods for the degradation of chemical carcinogens, with the aim of producing a compilation of acceptable techniques. It soon became evident that the published methods only dealt with the disappearance of the parent carcinogen and not with the products of degradation, which were sometimes mutagenic/carcinogenic (see Appendix). Some published methods gave products which could regenerate the original carcinogen on further treatment. It was decided, therefore, to change the original objectives, and to develop methods suitable for the degradation of chemical carcinogens. A joint meeting was organized in 1979,[45] and the following criteria for the validation of such methods were agreed:

• The method should afford efficient degradation of the compounds, as detected by conventional sensitive, analytical techniques.

- The residues produced should be non-mutagenic when tested *in vitro* in the Ames mutation assay,[46-47] using several strains of *Salmonella typhimurium*, with and without metabolic activation. It was recognized that long-term testing *in vivo* would provide better assurance of the absence of adverse biological effects of the residues, but such tests, if implemented, would slow down the development of degradation methods and are financially prohibitive.

- The method should be applicable in all laboratories, and should give comparable results. To meet this requirement, all the methods would be drafted in an International Organization for Standardization format, then subjected to collaborative study, and the results reviewed by the group of scientists who took part in these studies, with the object of accepting, modifying or discarding the methods.

In order to ensure success in the use of such methods it was recognized that:

i)   They should use inexpensive reagents;

ii) They should not be time consuming.

These two additional criteria were used when selecting potential methods for degradation.

Validated methods have been established for the chemical degradation of selected compounds from eight classes of chemical carcinogens: aflatoxins and other mycotoxins; nitrosamines; nitrosamides; polycyclic aromatic hydrocarbons; polycyclic heterocyclic hydrocarbons; hydrazines; haloethers; and aromatic amines, and a series of antineoplastic agents. These methods are described in detail as an Appendix to this chapter. During the development of the methods, several ways in which carcinogens might be encountered were considered, for example, solutions of the carcinogens in various solvents including aqueous solutions and the treatment of equipment and spillages.

Other investigations which have not yet been validated by collaborative studies are also included in the Appendix: the degradation of alkylating agents; ethidium bromide; chromium(VI); azo- and azoxy-compounds; and some halogenated compounds.

3 Summary of Available Options for the Disposal of Carcinogens

In the preceding sections various possibilities for the disposal of chemical carcinogens have been reviewed.

Evaporation or disposal of carcinogenic substances through the sewage system is ethically unacceptable and should be prohibited.

It has been demonstrated that landfill (storage or burial), even if carefully supervised, may be dangerous due to leakage, and could harm both the environment and those personnel who have to transport the material to the landfill site.

Incineration, when performed under rigorously controlled conditions, is suitable for the destruction of chemical carcinogens and wastes. Incinerators with adequate combustion capacities are, however, very expensive, both to acquire and to run, and only a few organizations can afford this method of disposal. There is also the difficulty, as with landfill storage or burial, of the handling and transport of genotoxic compounds. Incineration does not provide an answer to the problem of decontaminating spillages and reusable equipment. If equipment is not decontaminated, 'unexposed personnel' may be endangered, since they will most probably take less care in handling the equipment.[4] By contrast, chemical decontamination methods are not only useful for the *in situ* treatment of small quantities of contaminated wastes (thus offering a cheap alternative for laboratories handling small quantities of chemical carcinogens), but are also the only suitable techniques for the treatment of spillages and the decontamination of equipment.

4 References

1. International Agency for Research on Cancer: 'IARC Monographs on the Evaluation of Carcinogenic Risks to Humans: Overall Evaluation of Carcinogenicity; An Updating of IARC Monographs', Volumes 1-42, Supplement 7, IARC, Lyon, 1987.

2. International Agency for Research on Cancer: 'IARC Monographs on the Evaluation of Carcinogenic Risks to Humans: List of IARC Evaluations', IARC, Lyon, 1993.

3. D.K. Mayer, *Cancer*, 1992, **70**, 988.

4. P.J.M. Sessink, K.A. Boer, A.P.H. Sheefhals, R.B.M. Anzion, and R.P. Bos, *Int. Arch. Occup. Environ. Health*, 1992, **64**, 105.

5. R.M. Joyce, *Science*, 1984, **224**, 449.

6. Committee on Hazardous Substances in the Laboratory; Commission on Physical Sciences, Mathematics and Resources; National Research Council, 'Prudent Practices for Disposal of Chemicals from Laboratories', National Academy Press, Washington, DC, 1983, p. 15.

7. Journal Officiel de la République Française, Loi du 15 Juillet 1975, No. 75633, Modifiée par loi du 15 Décembre 1988, No. 881261.

8. Journal Officiel de la République Française, Loi du 13 Juillet 1992, No. 92646.

9. 'Safety and Accident Prevention in Chemical Operations', ed. H.H. Fawcett and W.S. Wood, J. Wiley and Sons, Interscience, New York, 1965.

10. J.L. Dwyer, 'Contamination Analysis and Control', Reinhold, New York, 1966.

11 'Handbook of Laboratory Safety', ed. N.V. Steere, Chemical Rubber Co., Cleveland, Ohio, 1971.

12. Manufacturing Chemists Association, 'Laboratory Waste Disposal Manual', Washington, DC, 1973.

13. American Institute of Chemical Engineers, 'Control of Hazardous Material Spills', Washington, DC, 1974.

14. American Chemical Society, 'Safety in Academic Chemistry Laboratories', Washington, DC, 1976.

15. M.A. Armour, L.M. Browne, and G.L. Weir, 'Hazardous Chemicals Information and Disposal Guide', University of Alberta, 1984.

16. G. Lunn, M. Castegnaro, and E.B. Sansone, 'Chemical Induction of Cancer', Vol. IIIC, Academic Press, San Diego, 1988.

17. A. Picot and P. Grenouillet, 'La Securité en Laboratoire de Chimie et de Biochimie', Lavoisier Tec. Doc., Paris, 1992.

18. M. Castegnaro and X. Rousselin, 'Cancérogènes, Mutagénes Chimiques et Autres Substances Toxiques: Traitement des Déchets avant Rejets'. INRS, Paris, 1993.

19. H.P. Hynes, in 'Environmental Sampling for Hazardous Wastes', ed. G.E. Schweitzer and J.A. Santolucito, ACS Symposium Series 267, American Chemical Society, Washington, DC, 1984, p. 1.

20. D.W. Miller, in 'Assessment of Health Effects at Chemical Disposal Sites', ed. W.W. Lowrance, The Rockefeller University, New York, 1981, p. 23.

21. OSHA, *Am. J. Hosp. Pharm.*, 1986, **43**, 1193.

22. C.S. Kim, R. Narang, A. Richards, K. Aldous, P. O'Keefe, R. Smith, D. Hilker, B. Bush, J. Slack, and D.W. Owens, in 'Hazardous Waste Disposal', ed. J.H. Highland, Ann Arbor Science, Michigan, 1980, p. 77.

23. K. Yamada, J. Fukuyama, and A. Honda, *Seikaku Eisei*, 1981, **25**, 78.

24. R.J. Lewis, Sr., 'Sax's Dangerous Properties of Industrial Materials', 8th Edition, Van Nostrand Reinhold, New York, 1992, p. 2783.

25. The International Technical Information Institute, 'Toxic and Hazardous Industrial Chemicals Safety Manual', 1976, p. 413.

26. M. Omura, T. Inanasu, and N. Ishinishi, *Mutat. Res.*, 1992, **298**, 125.

27. 'Alabama and EPA to examine company's waste disposal plan', New York Times, December 28, 1980.

28. 'Europe's plan for coping with toxic wastes', *Chem. Week*, March 2, 1983, p. 76.

29. D. Dickson, *Science*, 1983, **220**, 1362.

30. 'Falsch Deklarierte Sendung mit Dioxin aus Linz', Neue Zürcher Zeitung, July 30, 1983.

31 UNEP, 'Managing hazardous wastes', Newsletter of the Basel Convention, May 1992, p. 6.

32. UNEP, 'The Basel Convention on the Control of Transboundary Movements of Hazardous Wastes and their Disposal', UNEP Environmental Law Library, No. 2, 1990.

33. K.E. Kelly, in 'Hazardous Waste and Hazardous Materials', Vol. 3, Mary Ann Liebert Inc., Publishers, 1986, p. 367.

34. A. Levin, D.B. Fratt, A. Leonard, R.J.F. Bruins, and L. Fradkin, *J. Air Waste Manage. Assoc.*, 1991, **41**, 20.

35. B.S. Shane, C.B. Henry, J.H. Hotchkiss, K.A. Klausner, W.H. Gutenmann, and D.J. Lisk, *Arch. Environ. Contam. Toxicol.*, 1990, **19**, 665.

36. N. Steisel, R. Morris, and M.J. Clarke, *Waste Manage. Res.*, 1987, **5**, 381.

37. P.H. Brunner and H. Mönch, *Waste Manage. Res.*, 1986, **4**, 105.

38. M.S. Barbeito, *ACS Symp. Ser.*, 1979, **79**, 191.

39. W.A. Rubey, D.L. Hall, J.L. Tores, B. Dellinger, and R.A. Carnes, in 'Proceedings of the 7th International Symposium on Polynuclear Aromatic Hydrocarbons', Springer Verlag, New York, 1983, p. 1047.

40. C. Rappe, in 'Dioxins: Toxicological and Chemical Aspects', ed. F. Cattabeni, A. Cavallaro, and G. Galli, J. Wiley, 1978.

41. T.K. Wilkinson and H.W. Rogers, in 'Safe Handling of Chemical Carcinogens, Mutagens, Teratogens, and Highly Toxic Substances', ed. D.B. Walters, Ann Arbor Science, Michigan, 1980, p. 575.

42. P.T. Chien and M.H. Thomas, *J. Environ. Pathol. Toxicol.*, 1978, **2**, 513.

43. M.M. Coombs and M. Castegnaro, in 'The Disposal of Hazardous Waste from Laboratories', The Royal Society of Chemistry, London, 1983, p. 31.

44. International Agency for Research on Cancer, 'Handling of Chemical Carcinogens in the Laboratory: Problems of Safety', ed. R. Montesano, H. Bartsch, E. Boyland, G. Della Porta, L. Fishbein, R.A. Griesemer, A.B. Swan, and L. Tomatis, *IARC Scientific Publications* No. 33, IARC, Lyon, 1979.

45. International Agency for Research on Cancer, Internal Technical Report No. 79/002, 1979.

46. B.N. Ames, J. McCann, and Y. Yamasaki, *Mutat. Res.*, 1975, **31**, 347.

47. H. Bartsch, C. Malaveille, A.M. Camus, G. Martel-Planche, G. Brun, A. Hautefeuille, N. Sabadie, A. Barbin, T. Kuroki, C. Drevon, C. Piccoli, and R. Montesano, *Mutat. Res.*, 1980, **76**, 1.

5 Appendix — Experimental Details for the Degradation of Carcinogens

5.1 Introduction

This Appendix contains experimental details for the conversion of various classes of carcinogenic and mutagenic compounds into non-mutagenic residues. The results of some unsuccessful degradation methods are also presented. Literature references are included with the experimental methods in each section, and should be consulted for further details — in particular for the applicability of a method to a specific spillage or disposal problem, and for analytical techniques used to check degradation.

It must be stressed that if a method has been successful even for a number of compounds from one family, it may not be efficient for every congener. It is advisable, therefore, to investigate the efficiency of a method for each new compound.

Several classes of compounds can be degraded by oxidation with potassium permanganate under acidic conditions. This method was found to produce mutagenic Mn(II) which can be rendered non-mutagenic by conversion into Mn(OH)$_2$ with alkali, which rapidly darkens by air oxidation, before filtration and disposal of the residual mixture.

Aqueous solutions or mixtures of acidic, alkaline, reductive, or oxidative reagents are used in the procedures described below, unless another solvent is specified.

Ames tests (with and without metabolic activation) were carried out on degraded reaction mixtures, and, concurrently, with the appropriate positive controls. Negative controls (the degradation medium in the absence of the compound to be degraded) were also checked. In each case at least four concentrations were tested, and a response was considered positive when the mutation level was twice the spontaneous background of mutants and there was also a dose-related response.

5.2 Aflatoxins and Other Mycotoxins

5.2.1 Compounds Investigated

Aflatoxin B$_1$, aflatoxin B$_2$, aflatoxin G$_1$, aflatoxin G$_2$, aflatoxin M$_1$, citrinin, ochratoxin (OA), patulin, sterigmatocystin.

5.2.2 Methods Tested but Not Recommended for Use

- Treatment of aflatoxins with aqueous sodium hydroxide or sodium bicarbonate (the aflatoxin may be reformed on acidification).

- Treatment of aflatoxin B$_1$ with sodium hypochlorite which leads to the formation of a carcinogenic 2,3-dichloro derivative.

- Treatment with a strong mixture of chromic and sulfuric acids which, although effective, uses a carcinogen, Cr(VI).

- Ammoniation of ochratoxin A (OA) at room temperature or on warming (because OA can be reformed upon neutralization).

- Heat treatment of OA in aqueous solution.

- Degradation of sterigmatocystin with 6 mol l^{-1} sulfuric acid which leads to formation of mutagenic residues.

5.2.3 Validated Methods

Method 1: Aflatoxins
20 ml of sodium hypochlorite solution (5% available chlorine) are sufficient to degrade 20 µg of pure aflatoxins. Other components in the waste may also react with sodium hypochlorite. It is recommended, therefore, that the efficiency of the degradation of the aflatoxins is checked. To ensure effective degradation, an adequate excess of sodium hypochlorite should be used, *i.e.* a minimum of twice the amount estimated to be necessary.

Method 2: Aflatoxins and Patulin in Litter
Spread the litter contaminated by aflatoxins or patulin to a maximum depth of 5 cm and sprinkle it with 5% ammonia solution (16 mg/10 g litter). Autoclave for 20 min. at 128-130 °C.

Method 3: Aflatoxins
10 ml of 0.1 mol l^{-1} potassium permanganate in 1 mol l^{-1} sulfuric acid will degrade a mixture of 20 µg of aflatoxins in 3 h. Other components in the waste may react with potassium permanganate (turning the purple colour to brown). In such cases, it is recommended that the residual solution is analysed to confirm complete degradation.

Method 4: Carcasses Contaminated with Aflatoxins
Place the carcass on a shallow bed of quicklime and cover with quicklime to a depth of about 1 cm.

Method 5: Citrinin and OA
Solid citrinin (400 µg) or a solution of 400 µg OA or citrinin in 400 µl of ethanol (OA) or 800 µl of methanol (citrinin) is completely degraded by treatment with 10 ml of sodium hypochlorite solution (5% available chlorine) for 30 min. Since other compounds in the waste may also react with sodium hypochlorite, it is recommended that the efficiency of the degradation is checked using any sensitive analytical procedure.

Method 6: Citrinin
400 µg of citrinin are degraded by treatment overnight with 10 ml 10% ammonia at 100 °C.

Method 7: Sterigmatocystin
200 µg of sterigmatocystin in 4 ml of methanol are completely degraded by treatment with 5 ml of sodium hypochlorite solution (5% available chlorine) for 1 h. Further treatment with acetone removes any mutagenic compounds present.

Method 8: All Mycotoxins Cited Above
10 ml of 0.3 mol l^{-1} potassium permanganate in 2 mol l^{-1} sodium hydroxide will degrade 400 µg patulin, 300 µg of sterigmatocystin or AFB_1 or AFB_2 or AFG_1 or AFG_2, 2 mg citrinin or OA, in 3 h. Other components in waste may react with potassium permanganate turning the purple/green colour to brown.

5.2.4 *References*

• 'Laboratory Decontamination and Destruction of Aflatoxins B_1, B_2, G_1, G_2 in Laboratory Wastes', ed. M. Castegnaro, D.C. Hunt, E.B. Sansone, P.L. Schuller, M.G. Siriwardana, G.M. Telling, H.P. Van Egmond, and E.A. Walker, *IARC Scientific Publications* No. 37, Lyon, 1980.

• M. Castegnaro, M. Friesen, J. Michelon, and E.A. Walker, *Am. Ind. Hyg. Assoc. J.*, 1981, **42**, 398.

• 'Laboratory Decontamination and Destruction of Carcinogens in Laboratory Wastes: Some Mycotoxins', ed. M. Castegnaro, J. Barek, J.M. Fremy, M. Lafontaine, M. Miraglia, E.B. Sansone, and G.M. Telling, *IARC Scientific Publications* No. 113, 1991.

5.3 *N*-Nitrosamines

5.3.1 *Compounds Investigated*

N-Nitrosodimethylamine, *N*-nitrosodiethylamine, *N*-nitrosodipropylamine, *N*-nitrosodibutylamine, *N*-nitrosopiperidine, *N*-nitrosopyrrolidine (NPYR), *N*-nitrosomorpholine, *N-N'*-dinitrosopiperazine.

5.3.2 *Methods Tested but Not Recommended for Use*

- Treatment with sodium hypochlorite (inefficient).

- Treatment with a mixture of chromic and sulfuric acids (requires 5 days treatment and uses carcinogenic Cr(VI)).

- Reduction with CuCl/HCl (not reproducible).

- Oxidation by peroxotrifluoracetic acid (produces carcinogenic nitramines).

- Reduction by aluminium powder in alkaline medium (produces carcinogenic hydrazines).

5.3.3 *Validated Methods*

Method 1:
A solution of the *N*-nitrosamine in dichloromethane or another suitable solvent is concentrated, dried, and treated with an excess of hydrobromic acid solution (3%) on the basis that 5 ml of the HBr solution is sufficient to degrade 1 mg of *N*-nitrosamine in 1-2 ml of solvent within 15 min. NPYR is the exception to this, requiring 10 ml of hydrobromic acid solution to degrade 1 mg in 90 min.

The rate of reaction is drastically decreased by the presence of water or dimethyl sulfoxide (DMSO).

Method 2:
50 ml of potassium permanganate (0.3 mol l^{-1}) in sulfuric acid (3 mol l^{-1}) will degrade a mixture containing approx. 300 µg of *N*-nitrosamines. Other components in the waste may react with potassium permanganate (turning the purple colour to brown). Thus in all cases sufficient permanganate should be added to maintain a permanent purple colour.

Method 3: Chemical Apparatus Contaminated with N-*Nitrosamines*
Rinse and drain the apparatus 5 times using an appropriate volume of solvent, and combine the rinses. (See Table 1 for recommended decontamination solvents.)

Note: The volume used should be kept to a practical minimum bearing in mind that the solvent residues will themselves require decontamination.

Table 1 *Recommended solvents for removing* N-*nitrosamines from apparatus*

Type of contaminant	Recommended decontamination solvent		
	CH_2Cl_2	MeOH	H_2O
Undiluted *N*-nitrosamines	x		
Aqueous solutions			x
Alcoholic solutions	x	x	x
Oily solutions	x		
CH_2Cl_2	x		

Method 4:

50 g of nickel:aluminium alloy (50:50) are sufficient to destroy 5 g of *N*-nitrosamines in 1 l of 0.5 mol l⁻¹ potassium hydroxide. Other components in the waste may also react with the nickel-aluminium alloy or may poison the nickel catalyst which is formed. It is recommended, therefore, that the efficiency of the degradation is checked.

5.3.4 References

- R.T. Chien and M.H. Thomas, *J. Env. Path. Toxicol.*, 1978, **2**, 513.

- G.C. Emmet, C.I. Michejda, E.B. Sansone, and L.K. Keefer, in 'Safe Handling of Chemical Carcinogens, Mutagens, Teratogens and Highly Toxic Substances', ed. D.B. Walters, Ann Arbor Science, Michigan, 1979, p. 535.

- G. Lunn, E.B. Sansone, and L.K. Keefer, *Food. Chem. Toxicol.*, 1981, **19**, 493.

- 'Laboratory Decontamination and Destruction of Carcinogens in Laboratory Wastes: Some *N*-Nitrosamines', ed. M. Castegnaro, G. Eisenbrand, G. Ellen, L. Keefer, D. Klein, E.B. Sansone, D. Spincer, G.M. Telling, and K. Webb, *IARC Scientific Publications* No. 43, Lyon, 1982.

- M. Castegnaro, J. Michelon, and E.A. Walker, in '*N*-Nitroso Compounds: Occurrence and Biological Effects', ed. H. Bartsch, I.K. O'Neill, M. Castegnaro, and M. Okada, *IARC Scientific Publications* No. 41, 1982, p. 151.

- G. Lunn, E.B., Sansone, and L.K. Keefer, *Carcinogenesis*, 1983, **4**, 315.

5.4 Polycyclic Aromatic and Heterocyclic Compounds (PAC and PHC)

5.4.1 Compounds Investigated

Benzo[a]anthracene (BA), benzo[a]pyrene (BP), 7-bromomethylbenz[a]anthracene (BrMBA), dibenz[a,h]anthracene (DBA), 7,12-dimethylbenz[a,h]anthrazene (DMBA), 3-methylcholanthrene (MC), dibenz[a,j]acridine (DB[a,j]AC), dibenz[a,h]acridine (DB[a,h]AC), 7H-dibenzo[c,g]carbazole (DB[c,g]C), 13H-dibenzo[a,i]carbazole (DB[a,i]C).

5.4.2 Methods Tested but Not Recommended for Use

- Treatment of PAH with sodium hypochlorite (inefficient).

- Reduction of PAH by nickel:aluminium alloy (50:50) under alkaline conditions (inefficient).

- Treatment with a mixture of chromic and sulfuric acids, (effective but uses carcinogenic Cr(VI)).

- Oxidation of dibenz[a,h]anthracene and dibenz[a,h]acridine with a saturated solution of potassium permanganate (degradation incomplete).

- Oxidation of dibenz[a,h]acridine with alkaline potassium permanganate (degradation incomplete).

- Treatment of the two dibenzacridines with concentrated sulfuric acid (degradation incomplete).

5.4.3 Validated Methods

Method 1: PAH and PHC
10 ml of a solution containing 0.3 mol l^{-1} potassium permanganate in 3 mol l^{-1} sulfuric acid will degrade 5 mg of BA, BP, DMBA, DBA, MC, or BrMBA in acetone in 1 h. Other components in wastes may react with potassium permanganate, turning the purple colour to brown. It is recommended, therefore, that the efficiency of the degradation is checked. The four polycyclic heterocyclic compounds have also been successfully degraded by this method giving non-mutagenic residues, but the method has not been validated.

Method 2: Some PAH and PHC
10 ml of a saturated solution of potassium permanganate will degrade 1 mg of BA, BP, DMBA, MC, or BrMBA dissolved in 2 ml acetone, or 5 mg of DB[c,g]C or DB[a,i]C, or 1 mg of DB[a,j]AC in 2 ml of acetonitrile. Other components in the wastes may react with potassium permanganate, turning the purple colour to brown. Under such circumstances, it is recommended that the residual solution should be analysed for completeness of degradation.

Method 3: PAH and Some PHC
10 ml concentrated sulfuric acid will degrade 5 mg BA, BP, DMBA, DBA, MC, or BrMBA, or DB[c,g]C, or DB[a,i]C dissolved in 2 ml DMSO in 2 h.

Note 1: The efficiency of destruction depends upon the ratio of sulfuric acid: DMSO; this should not be less than 5:1.

Note 2: For the compounds tested, 2 h was found to be a satisfactory reaction time. When using this method to degrade other PAH or PHC, it is strongly recommended that a reaction time sufficient for complete destruction is established beforehand.

Note 3: The reaction has been tested using other solvents, such as acetone and DMF. For most of the compounds tested, the reaction was found to proceed satisfactorily, but longer reaction periods were required.

Method 4: PHC
5 mg of DB[c,g]C, DB[a,i]C, DB[a,j]AC, or DB[a,h]AC in 5 ml of acetone are completely degraded by treatment for about 1 h with 0.2 to 0.3 g iron(II) chloride and 10 ml H_2O_2.

Method 5: Some PHC
5 mg of DB[c,g]C, or DB[a,i]C, or DB[a,j]AC in 2 ml of acetonitrile are completely degraded to non-mutagenic products by treatment for 3 h with 10 ml of a solution containing 0.3 mol l^{-1} potassium permanganate in 2 mol l^{-1} sodium hydroxide.

Note 1: Under these conditions 0.5 mg of DB[a,h]AC is degraded only to the extent of 76% and this method is not recommended.

Note 2: Other components in wastes may react with potassium permanganate, turning the purple/green colour to brown. The efficiency of degradation should be checked using suitable analytical procedures.

5.4.4 References

• M. Castegnaro, M. Coombs, M.A. Phillipson, M.C. Bourgade, and J. Michelon, in 'Proceedings of the 7th Symposium on Polynuclear Hydrocarbons', Springer Verlag, New York, 1983, p. 257.

• 'Laboratory Decontamination and Destruction of Carcinogens in Laboratory Wastes: Some Polycyclic Aromatic Hydrocarbons', ed. M. Castegnaro G. Grimmer, O. Hutzinger, W. Karcher, H. Kunte, M. Lafontaine, E.B. Sansone, G. Telling, and S.P. Tucker, *IARC Scientific Publications* No. 49, Lyon, 1983.

• 'Laboratory Decontamination and Destruction of Carcinogens in Laboratory Wastes: Some Polycyclic Heterocyclic Compounds', ed. M. Castegnaro, J. Barek, J. Jacob, U. Kirso, M. Lafontaine, E.B. Sansone, G.M. Telling, and T. Vu Duc, *IARC Scientific Publications* No. 114, Lyon, 1991.

• G. Lunn, E.B. Sansone, M. De Méo, M. Laget, and M. Castegnaro, *Am. Ind. Hyg. Assoc. J.*, 1993, (in press).

5.5 Nitrosamides

5.5.1 Compounds Investigated

N-nitroso-*N*-methylurea (MNU), *N*-nitroso-*N*-ethylurea (ENU), *N*-nitroso-*N*-methylurethane (MNUT), *N*-nitroso-*N*-ethylurethane (ENUT), *N*-nitroso-*N*'-nitro-*N*-methylguanidine (MNNG).

5.5.2 Methods Tested but Not Recommended for Use

• Treatment with an alkaline reagent (a diazoalkane may be generated).

• Oxidation with potassium permanganate under acidic conditions for 0.5 h (produces a mutagenic residue).

• Treatment with dilute hydrochloric acid (3 mol l^{-1}) in presence of iron powder and a medium containing acetone (generates mutagenic residues).

5.5.3 Validated Methods

Method 1:
35 g of sulfamic acid in 1 l of 3 mol l^{-1} hydrochloric acid are sufficient to destroy 17 g N-nitrosamides. Other components in waste may interfere with the destruction process, so it is recommended that the efficiency of the degradation is checked.

Note 1: This method produces mutagenic residues with ENUT and must not be used for this compound.

Note 2: The method has also been successfully tested on N-nitroso-N-methyl-p-toluenesulfonamide (MNTS) and N-nitroso-N'-nitro-N-ethylguanidine (ENNG) but not validated.

Method 2:
35 g of iron filings in 1 l of solution containing 3 mol l^{-1} hydrochloric acid are sufficient to destroy 17 g N-nitrosamides. Other components in waste may interfer with the destruction process, so it is recommended that the efficiency of the degradation is checked.

Note 1: This method must not be used in presence of acetone (see Section 5.5.2).

Method 3:
10 ml of a solution containing 0.3 mol l^{-1} potassium permanganate in 3 mol l^{-1} sulfuric acid will degrade 50 mg of any of the following nitrosamides: MNU, ENU, MNNG, MNUT, or ENUT, within 8 h. Although the actual chemical degradation takes much less time, it is necessary to allow reaction to proceed for 8 h to obtain non-mutagenic residues.

Note: This method must not be used for shorter periods (see Section 5.5.2).

Method 4:
A solution of 100 mg N-nitrosamides in 2 ml dry dichloromethane, ethyl acetate or any other suitable solvent is concentrated and treated with 10 ml of a solution of 3% hydrobromic acid to give quantitative degradation of N-nitrosamides within 15 min. The NOBr formed is removed by flushing with nitrogen for 30 min.

Note: The rate of reaction is drastically reduced by the presence of water. Alcohols also retard the reaction, and longer reaction periods are necessary.

5.5.4 *References*

* 'Laboratory Decontamination and Destruction of Carcinogens in Laboratory Wastes: Some *N*-Nitrosamides', ed. M. Castegnaro, M. Bernard, L.W. Van Broekhoven, D. Fine, R. Massey, E.B. Sansone, P.L.R. Smith, B. Spiegelhalder, A. Stacchini, and J.J. Vallon. *IARC Scientific Publications* No. 55, Lyon, 1984.

* G. Lunn, E.B. Sansone, A.W. Andrews, M. Castegnaro, C. Malaveille, J. Michelon, I. Brouet, and L.K. Keefer, in '*N*-Nitroso Compounds: Occurrence, Biological Effects and Relevance to Human Cancer.' ed. I.K. O'Neill, R.C. Von Borstel, C.T. Miller, J. Long, and H. Bartsch, *IARC Scientific Publications* No. 57, Lyon, 1984.

* G. Lunn and E.B. Sansone, *Food. Chem. Toxicol.*, 1988, **26**, 481.

5.6 Hydrazines

5.6.1 *Compounds Investigated*

Hydrazine, methylhydrazine, 1,1-dimethylhydrazine, 1,2-dimethylhydrazine, procarbazine.

5.6.2 *Methods Tested but Not Recommended for Use*

* Oxidation by potassium permanganate in an acidic solution of the asymmetrical dialkyl- or diaryl-hydrazines (the corresponding carcinogenic *N*-nitrosamines are formed).

* Oxidation by potassium permanganate in acid conditions and in presence of DMSO. With low levels of DMSO, small quantities of nitrosamines are produced, and high levels of DMSO inhibit the oxidation.

* Oxidation by stoichiometric quantities of calcium hypochlorite for 30 min (mutagenic residues are produced).

* Oxidation of 1,1-dimethylhydrazine by sodium hypochlorite (a small fraction is converted to *N*-nitrosodimethylamine).

5.6.3 Validated Methods

Method 1:
50 g of nickel:aluminium alloy (50:50) are sufficient to destroy 5 g of hydrazines in 1 l of 0.5 mol l^{-1} potassium hydroxide. Other compounds in the waste may also react with the nickel-aluminium alloy or may poison the nickel catalyst which is formed. It is recommended, therefore, that the efficiency of the degradation is checked.

Note 1: Addition of nickel-aluminium alloy to an alkaline solution results in a highly exothermic reaction with the evolution of large quantities of hydrogen. It is essential to add the alloy slowly over a period of time while cooling the reaction vessel in an ice bath.

Note 2: Care should be taken that the nickel-aluminium alloy powder is kept in suspension throughout the operation (avoid lumps or adherence to sides of the reaction vessel).

Method 2:
5 ml of a solution containing 0.3 mol l^{-1} potassium permanganate and 3 mol l^{-1} sulfuric acid can degrade 25 mg aliquots of the following symmetrical hydrazines: MMH, 1,2-dimethylhydrazine, procarbazine or hydrazine, in less than 30 min. The reaction mixtures should be allowed to stand overnight to remove any traces of nitrosamines that may have formed.

Note: *cf.* 5.6.2.

Method 3:
In about 4 mol l^{-1} hydrochloric acid, 1 mol hydrazine or 1 mol 1,2-dimethylhydrazine is oxidized by 1 mol potassium iodate.

Note: The residues from this degradation were toxic, and the unequivocal absence of mutagenic activity could not therefore be demonstrated.

Method 4:
Although 1.5 g calcium hypochlorite can degrade 100 mg of each of the hydrazine compounds investigated within 30 min, it was found necessary to quadruple the amount of oxidant and to lengthen the reaction time to 12 h to obtain non-mutagenic residues.

5.6.4 References

- 'Laboratory Decontamination and Destruction of Carcinogens in Laboratory Wastes: Some Hydrazines', ed. M. Castegnaro, G. Ellen, M. Lafontaine, H.C. Van der Plas, E.B. Sansone, and S.P. Tucker, *IARC Scientific Publications* No. 54, Lyon, 1983.

- G. Lunn, E.B. Sansone, and L.K. Keefer, *Environ. Sci. Technol.*, 1983, **17**, 240.

5.7 Haloethers

5.7.1 Compounds Investigated

Chloromethylmethyl ether (CMME) and bis(chloromethyl) ether (BCME).

5.7.2 Validated Methods

Method 1
In water-miscible solvents, 1 ml of 6% ammonia will degrade 50 mg of CMME and/or BCME in 3 h.
   In water-immiscible solvents, 1 ml of 33% ammonia will degrade 50 mg of CMME and/or BCME in 3 h.

Note: The reaction proceeds rapidly for CMME, but since this compound always contains BCME as an impurity, it is necessary to allow 3 h for completion of the degradation reaction.

Method 2:
3.5 ml of 15% m/v sodium phenate in methanol will degrade 50 mg of either CMME or BCME in 1 ml of solvent in 3 h.

Note: The reaction proceeds rapidly for CMME, but since this compound always contains BCME as an impurity, it is necessary to allow 3 h for completion of the degradation reaction.

Method 3:
1.5 ml of a 8-9% m/v sodium methoxide in methanol will degrade 50 mg of either CMME or BCME in 1 ml of solvent in 3 h.

Note: The reaction proceeds rapidly for CMME, but since this compound always contains BCME as an impurity, it is necessary to allow 3 h for completion of the degradation reaction.

5.7.3 References

- M. Alvarez and R.T. Rosen. *Int. J. Environ. Anal. Chem.*, 1976, **4**, 241.

- 'Laboratory Decontamination and Destruction of Carcinogens in Laboratory Wastes: Some Haloethers', ed. M. Castegnaro, M. Alvarez, M. Iovu, E.B. Sansone, G.M. Telling, and D.T. Williams, *IARC Scientific Publications* No. 61, Lyon, 1984.

5.8 Aromatic Amines and 4-Nitrobiphenyl

5.8.1 Compounds Investigated

4-Aminobiphenyl (4-ABP), benzidine (Bz), 3,3'-dichlorobenzidine (DClB), 3,3'-dimethoxybenzidine (DMoB), 3,3'-dimethylbenzidine (DMB), 4,4'-methylenebis-(2-chloroaniline) (MOCA), 1-naphthylamine (1-NAP), 2-naphthylamine (2-NAP), 2,4-diaminotoluene (TOL), 4-nitrobiphenyl (4-NBP), 3,3'-diaminobenzidine (DAB).

5.8.2 Methods Tested but Not Recommended for Use

- Reduction of 4-nitrobiphenyl by nickel:aluminium (50:50) in alkaline medium followed by treatment of the resulting 4-ABP (yield of 4-ABP is poor).

- Deamination of 1-NAP, 2-NAP, and TOL by diazotization in presence of hypophosphorous acid (mutagenic residues are obtained).

- Diazotization of DClB, DMoB, 1-NAP, and TOL (mutagenic residues are obtained).

- Treatment of DAB with sodium hypochlorite (produces mutagenic residues).

5.8.3 Validated Methods

Method 1:
Bz, DClB, DMB, DMoB, TOL, 1-NAP, and 2-NAP dissolved in 10 ml of 0.1 mol l^{-1} hydrochloric acid at a concentration of 0.005 mol l^{-1}; or MOCA dissolved in 1 mol l^{-1} sulfuric acid at a concentration of 0.001 mol l^{-1}; or 4-ABP dissolved in glacial acetic acid at a concentration of 0.001 mol l^{-1}, are degraded by the action of 5 ml potassium permanganate (0.2 mol l^{-1}) and 5 ml sulfuric acid (2 mol l^{-1}) within 10 h.

Note: This method has also been used for the degradation of DAB but has not been validated.

Method 2:
Up to 100 mg of the amines listed in 5.8.1, when dissolved in 1 l of 1 g l^{-1} sodium acetate containing up to 20% ethanol, are degraded by 0.003 mol l^{-1} hydrogen peroxide in the presence of 300 units of peroxidase within 3 h. This method is not suitable for 4-ABP.

Note: This method produces solid residues, which possess mutagenic activity and must be retreated.

Method 3:
The amines, dissolved in 5 ml 10% hypophosphorous acid at a concentration of 0.025 mol l^{-1}, are treated with 5 ml sodium nitrite (0.5 mol l^{-1}).

Note 1: See 5.8.2 for exclusions.

Note 2: Slightly mutagenic residues have been obtained by this method for Bz and its derivatives. It should be used with caution for these compounds.

Note 3: DAB is not completely degraded.

Method 4:
4-NBP dissolved in 10 ml glacial acetic acid at a concentration of 0.005 mol l^{-1} is reduced to 4-ABP by zinc powder (50 times molar excess) in the presence of 10 ml sulfuric acid (2 mol l^{-1}) while stirring overnight. The resulting 4-ABP is degraded by the action of 10 ml potassium permanganate (0.2 mol l^{-1}) within 10 h.

Method 5
250 mg of MOCA, Bz, DMB, TOL, dissolved in 10 ml water followed by the addition of 10 ml concentrated sulphuric acid, while stirring, are diazotized at 0 °C by the addition of 250 mg sodium nitrite in 5 ml water. The reaction is then allowed to proceed while stirring for 20 min. at room temperature and then 20 min. at *ca.* 90 °C. A 10 mol l^{-1} sodium hydroxide solution is added to neutralize the mixture.

Note: The method is not suitable for DClB, DMoB 1-NAP and TOL

5.8.4 References

• A.M. Klibanov, E.D. Morris, *Enz. Microbiol. Technol.*, 1981, **3**, 119.

• 'Laboratory Decontamination and Destruction of Carcinogens in Laboratory Wastes: Some Aromatic Amines and 4-Nitrobiphenyl', ed. M. Castegnaro, J. Barek, J. Denis, G. Ellen, M. Klibanov, M. Lafontaine, R. Mitchum, P. Van Roosmalen, E.B. Sansone, L.A. Sternson, and M. Vahl, *IARC Scientific Publications* No. 64, Lyon, 1985.

- M. Castegnaro, C. Malaveille, I. Brouet, J. Michelon, and J. Barek, *Am. Ind. Hyg. Assoc. J.*, 1985, **46**, 187.

- G. Lunn and E.B. Sansone, *Appl. Occup. Environ. Hyg.*, 1991, **6**, 49.

5.9 Antineoplastic Agents and Other Drugs

5.9.1 Compounds Investigated

Doxorubicine, daunorubicine, methotrexate, dichloromethotrexate, cyclophosphamide, ifosfamide, vincristine sulfate, vinblastine sulfate, 6-thioguanine, 6-mercaptopurine, cisplatin, streptozotocin, chlorozotocin, lomustine, carmustine, PCNU, semustine, melphalan, dacarbazine, uracil mustard, procarbazine, spiromustine, isoniazide, iproniazide, mechlorethamine.

5.9.2 Methods Tested but Not Recommended For Use

Mutagenic residues were obtained from the following degradations:

- Doxorubicine and daunorubicine: oxidation by 5% or 10% sodium hypochlorite solution.

- Cyclophosphamide and ifosfamide: oxidation by potassium permanganate in acidic conditions.

- Cisplatin: oxidation by potassium permanganate in acidic conditions.

- Melphalan: oxidation by potassium permanganate in acidic conditions.

- Carmustine, semustine, PCNU: denitrosation by 4.5% HBr in glacial acetic acid.

- Lomustine, carmustine, semustine, PCNU, and chlorozotoxine: oxidation by potassium permanganate in acidic conditions.

- Dacarbazine: oxidation by potassium permanganate in acidic conditions.

5.9.3 Validated Methods

Method 1: Doxorubicine and daunorubicin
30 mg of doxorubicine or daunorubicin dissolved in 10 ml of 3 mol l⁻¹ sulfuric acid are degraded by 1 g potassium permanganate during 2 h.

Note: Very slight mutagenic activity has been detected from this degradation of doxorubicine.

Method 2: Methotrexate and dichloromethotrexate
50 mg of methotrexate or 10 mg of dichloromethotrexate (solid) dissolved in 10 ml of 3 mol l⁻¹ sulfuric acid are degraded by 0.5 g potassium permanganate in 1 h.

Note: In the case of pharmaceutical preparations containing dichloromethotrexate, up to 50 mg can be dissolved in 10 ml of 3 mol l⁻¹ sulfuric acid and degraded with 0.5 g of potassium permanganate.

Method 3: Methotrexate
50 mg of methotrexate, dissolved in 50 ml of 4% m/v sodium hydroxide, are degraded by 5.5 ml of 1% m/v potassium permanganate in 30 min.

Method 4: Methotrexate
50 mg of methotrexate dissolved in 100 ml of 4% m/v sodium hydroxide are degraded by 4.6 ml of 5.25% sodium hypochlorite in 30 min.

Method 5: Cyclophosphamide and ifosfamide
10 ml of 12% m/v sodium hydroxide are sufficient to degrade 100 mg cyclophosphamide or ifosfamide in 20 ml DMF, when refluxed for 4 h.

Method 6: Cyclophosphamide
A sample of 250 mg cyclophosphamide dissolved in 10 ml of 1 mol l⁻¹ hydrochloric acid is completely hydrolysed when refluxed for 1 h. After addition of 1.5 g sodium thiosulfate to the neutralized reaction mixture, the medium is made strongly alkaline with 20% m/v sodium hydroxide and the reaction allowed to proceed for 1 h.

Method 7: Vincristine and vinblastine sulfate
10 mg of vincristine sulfate or vinblastine sulfate in 10 ml of 3 mol l⁻¹ sulfuric acid are completely degraded by 0.5 g potassium permanganate in 2 h.

Method 8: 6-Thioguanine and 6-mercaptopurine
18 mg of 6-thioguanine or 6-mercaptopurine dissolved in 20 ml of 3 mol l⁻¹ sulfuric acid are degraded by 0.13 g potassium permanganate in 10-12 h.

Method 9: Cisplatin
30 mg of cisplatin dissolved in 50 ml of 2 mol l⁻¹ sulfuric acid are degraded by 1.5 g zinc powder in 10-12 h.

Method 10: Cisplatin
100 mg of cisplatin are degraded by 3 ml of 0.68 mol l⁻¹ sodium diethyldithio-carbamate solution in 0.1 mol l⁻¹ sodium hydroxide, followed by the addition of an equivalent amount of a saturated solution of sodium nitrate.

Method 11: Lomustine, chlorozotocin and streptozotocin
100 mg of lomustine dissolved in 2-3 ml dichloromethane, or 100 mg solid chlorozotocin or streptozotocin, are degraded by 10 ml of a 4.5% solution of hydrobromic acid in glacial acetic acid in 15 min. The nitrosyl bromide formed, is removed by flushing with nitrogen for 30 min. to prevent possible reformation of *N*-nitrosoureas.

Method 12: Streptozotocin
48 mg of streptozotocin dissolved in 10 ml of 3 mol l^{-1} sulfuric acid are degraded by 2 g potassium permanganate in 10-12 h.

Methods 13-15: Procarbazine (see Section 5.6.3).

5.9.4 Methods Not Validated

Method 16: Melphalan
20 mg of melphalan in 20 ml of 2 mol l^{-1} sodium hydroxide are degraded in 1 h by addition of 0.2 g of potassium permanganate.

Method 17: Dacarbazine
100 mg of dacarbazine in 10 ml of an aqueous solution containing 100 mg citric acid and 50 mg mannitol are completely degraded by addition of 10 ml of a 1 mol l^{-1} sodium hydroxide solution and 1 g of powdered nickel:aluminium (50:50) with stirring for 20 h.

Method 18: Lomustine, carmustine, chlorozotocine, semustine, PCNU, melphalan, uracil mustard, and spiromustine
Dissolve in methanol at about 10 mg ml^{-1}, add an equivalent amount of potassium hydroxide 2 mol l^{-1} and 1 g of nickel:aluminium (50:50) in powder form per 20 ml of solution. Allow to react overnight at room temperature.

Method 19: Mechlorethamine, chlorambucil, cyclophosphamide and ifosfamide
Dissolve in water at about 10 mg ml^{-1}, add an equivalent amount of 2 mol l^{-1} potassium hydroxide and 1 g of powdered nickel:aluminium (50:50) per 20 ml of solution. Allow to react overnight at room temperature.

Method 20: Isoniazide and iproniazide
Dissolve 100 mg of isoniazide or 50 mg of iproniazide in 10 ml of water, add an equivalent amount of 2 mol l^{-1} potassium hydroxide and 1 g of powdered nickel: aluminium (50:50) per 20 ml of solution. Allow to react overnight at room temperature.

Other Methods: Human urine contaminated with drugs and their metabolites
Methods for the treatment of human urine of patients treated with cisplatin, cyclophosphamide, doxorubicine, and mitomycin C are available (see references by Monteith *et al.* in Section 5.9.5).

5.9.5 *References*

- 'Laboratory Decontamination and Destruction of Carcinogens in Laboratory Wastes: Some Antineoplastic Agents', ed. M. Castegnaro, J. Adams, M.A. Armour, J. Barek, J. Benvenuto, C. Confalonieri, U. Goff, S. Ludeman, D. Reed, E.B. Sansone, and G. Telling, *IARC Scientific Publications* No. 73, Lyon, 1985.

- J. Barek, M. Castegnaro, C. Malaveille, I. Brouet, and J. Zima, *Microchemical J.*, 1987, **36**, 192.

- G. Lunn and E.B. Sansone, *Am. J. Hosp. Pharm.*, 1987, **44**, 2519.

- 'Laboratory Decontamination and Destruction of Carcinogens in Laboratory Wastes: Some Hydrazines', ed. M. Castegnaro, G. Ellen, M. Lafontaine, H.C. Van Der Plas, E.B. Sansone, and S.P. Tucker, *IARC Scientific Publication* No. 54, Lyon, 1983.

- G. Lunn, E.B. Sansone, and L.K. Keefer, *Environ. Sci. Technol.*, 1983, **17**, 240.

- G. Lunn, E.B. Sansone, A.W. Andrews, and L.C. Helling, *J. Pharm. Sci.*, 1989, **78**, 652.

- D.K. Monteith, T.H. Connor, J.A. Benvenuto, E.J. Fairchild, and J.C. Theiss, *Toxicol. Lett.*, 1988, **40**, 257.

- D.K. Monteith, T.H. Connor, J.A. Benvenuto, E.J. Fairchild, and J.C. Theiss, *Environ. Molec. Mutagen.*, 1987, **10**, 341.

5.10 Ethidium Bromide

5.10.1 *Methods Tested but Not Recommended*

- Treatment with sodium hypochlorite (mutagenic residues are formed).

- Treatment with nickel:aluminium alloy (50:50) in alkali (mutagenic residues are formed).

5.10.2 Methods Not Validated

Method 1:
1.5 g of ethidium bromide in 3 ml water are degraded on standing for 20 h following the addition of 600 µl of 5% hypophosphorous acid and 360 µl of 0.5 mol l^{-1} sodium nitrite.

Method 2:
Add 1 g charcoal per mg ethidium bromide to the solution to be decontaminated and allow to act at room temperature for at least 30 min. while shaking at least 3 times. Filter through fluted filter paper and incinerate or dispose of the paper and filtrate by suitable means.

Note 1: The efficiency of charcoal for absorbing ethidium bromide may vary from one batch to another by a factor of 10. Each batch should therefore be tested before use.

Note 2: This method has also been proposed for the decontamination of laboratory benches and other surfaces.

5.10.3 References

- P. Quillardet, C. De Bellecombe, M. Hoffnung. Le bromure d'éthidium perd-t-il son activité mutagène en présence d'eau de Javel? Poster présenté aux Journées de Biotechnologie de l'Institut Pasteur, Février 1987.

- G. Lunn and E.B. Sansone, *Anal. Biochem.*, 1987, **162**, 453.

- O. Bensaude, *Trends in Genetics*, 1988, **4**, 84.

- I. Muranyi-Kovacs, Le bromure d'éthidium: destruction et décontamination. *Dossier prévention*, numéro 2, INSERM, Avril 1988.

5.11 Some Alkylating Agents

5.11.1 Compounds Investigated

Dimethyl sulfate (DMS), diethyl sulfate (DES), methyl methanesulfonate (MMS), ethyl methanesulfonate (EMS).

5.11.2 Methods Not Validated

Method 1: DMS

i) 13.3 g of DMS in 500 ml of 1 mol l⁻¹ sodium hydroxide, *or* 1 mol l⁻¹ sodium carbonate, *or* 1.5 mol l⁻¹ ammonium hydroxide, are hydrolysed in 15 min.

ii) 0.1 ml of DMS in 1 ml methanol, ethanol, dimethylsufoxide (DMSO), acetone, or dimethylformanide (DMF) are hydrolysed by treatment with 4 ml of one of the above mentioned alkaline solutions for 15 min. (for methanol, ethanol, DMSO, or DMF solution), or 1 h for acetone solutions.

iii) 0.1 ml of DMS in 1 ml toluene, *p*-xylene, benzene, 1-pentanol, ethyl acetate, chloroform or carbon tetrachloride are degraded by shaking with 4 ml of one of the above mentioned alkaline solutions for 1 day.

Method 2: DMS, DES, MMS, and EMS

The compounds are degraded with 1 mol l⁻¹ $Na_2S_2O_3$ according to the kinetic formula $\log_e C = \log_e C_0 - at$ where C_0 is the initial concentration of alkylating agent, 'a' is a constant dependent on the compound to be degraded and the reaction temperature, and t is time (in minutes). At 25 °C, a = 4.85 (DMS), 0.73 (DES), 1.16 (MMS), 0.12 (EMS).

Thus 99.5% degradation is achieved in 1 min for DMS, 7 min (DES), 1.6 min (MMS), and 44 min (EMS).

5.11.3 References

• G. Lunn and E.B. Sansone, *Am. Ind. Hyg. Assoc. J.*, 1985, **46**, 3, 111.

• M. De Méo, M. Laget, M. Castegnaro, and G. Duménil, *Am. Ind. Hyg. Assoc. J.*, 1990, **51**, 505.

5.12 Chromium(VI)

5.12.1 Compounds Investigated

Sodium dichromate, potassium dichromate, chromium trioxide and chromic acid.

5.12.2 Methods Not Validated

Method 1: Solid compounds
Dissolve 5 g of the compound in 100 ml 0.5 mol l^{-1} sulfuric acid. Add 10 g sodium metabisulfite and allow to react while shaking the mixture for 1 h. Allow the mixture to reach room temperature and add 6 g magnesium hydroxide. Shake for 1 h, and allow to stand overnight.

Method 2: Solutions
Dilute 10 ml in 16 ml water and shake for 1 h. Add 10 ml of 10% m/v sodium metabisulfite solution and 12 g magnesium hydroxide. Shake for 1 h and allow to stand overnight.

5.12.3 Reference

- G. Lunn and E.B. Sansone, *J. Chem. Education*, 1989, **66**, 443.

5.13 Azo-, Azoxy-compounds and 2-Aminoanthracene

5.13.1 Compounds Investigated

Azobenzene, azoxyanisole, phenylazophenol, phenylazoaniline, Fast Garnet, 2-amino-anthracene.

5.13.2 Methods Not Validated

Method 1: Azobenzene, azoxyanisole, phenylazophenol, phenylazoaniline, 2-aminoanthracene, Fast Garnet
To 10 mg of compound (5 mg for phenylazoaniline) in 1 ml glacial acetic acid, add 40 ml (80 ml for phenylazoaniline) of 0.3 mol l^{-1} potassium permanganate in 3 mol l^{-1} sulfuric acid. Stir at room temperature for 18 h.

Method 2: Azobenzene, azoxybenzene, azoxyanisole, phenylazophenol
To a 5 mg ml^{-1} solution in methanol, add an equal volume of 2 mol l^{-1} potassium hydroxide and 1 g of nickel:aluminium alloy (50:50) per 20 ml of mixture. Allow to react while stirring at room temperature for 18 h.

Method 3: All compounds
Stir 10 ml of a saturated aqueous solution with 0.5 g amberlite XAD-16 for 18 h and then filter.

5.13.3 Reference

- G. Lunn and E.B. Sansone, *Appl. Occup. Environ. Hyg.*, 1991, **6**, 1020.

6 Acknowledgements

The author wishes to thank Mrs. Z. Schneider for secretarial assistance, the Office of Safety of the NIH and the French Ministry of the Environment, for their financial support towards the development of chemical methods for degradation of chemical carcinogens and J. Adams, University of Texas System Cancer Center, M.D. Anderson Hospital and Tumor Institute, Houston, USA; M. Alvarez, Chemical Research and Development Center, FMC Corporation, Princeton, USA; M.A. Armour, Department of Chemistry, University of Alberta, Edmonton, Canada; J. Barek, Department of Analytical Chemistry, Charles University, Prague, Czechoslovakia; M. Bernard, Ecole National Supérieure des Industries Agrocoles et Alimentaires, Douai, France; J. Benvenuto, Department of Pharmacy and Chemotherapy Research, University of Texas System Cancer Center, M.D. Anderson Hospital and Tumor Institute, Houston, USA; L. Van Broekhoven, Center for Agrobiological Research, Wageningen, The Netherlands; C. Confalonieri, Pharmaceutical Research Development, Farmitalia Carlo Erba, Milan, Italy; J. Dennis, Ministry of Agriculture, Fisheries and Food, Norwich, England; G. Eisenbrand (present address), Department of Chemistry and Environmental Toxicology, University of Kaiserslautern, Germany; H.P. Van Egmond, National Institute of Public Health, Bilthoven, The Netherlands; G. Ellen, National Institute of Public Health, Bilthoven, The Netherlands; D. Fine, Thermedics Inc., Woburn, MA, USA; J.M. Frèmy, Ministère de l'Agriculture, Paris, France; U. Goff, Thermedics Inc., Woburn, MA, USA; G. Grimmer, Biochemisches Institut für Umweltcarcinogene, Ahrensburg/Holst, Germany; D.G. Hunt, Laboratory of the Government Chemist, London, England; O. Hutzinger (present address), Chair of Ecological Chemistry and Geochemistry, University of Bayreuth, Germany; M. Iovu, Chimie Organica, Facultatea de Farmacie, Bucharest, Romania; J. Jacob, Biochemisches Institut für Umweltcarcinogene, Ahrensburg, Germany; W. Karcher, Petten Establishment, Joint Research Center, Commission of the European Communities, Petten, The Netherlands; L.K. Keefer (present address), Laboratory of Comparative Carcinogenesis, Division of Cancer Etiology, NCI, Frederick, USA; U. Kirso, Institute of Chemistry, Estonian Academy of Sciences, Tallinn, Estonia; D. Klein (present address), Centre de Recherche Creealis, Brive La Gaillarde, France; M. Klibanov, Department of Nutrition and Food Science, Massachusetts Institute of Technology, Cambridge, USA; H. Kunte, Hygiene Institut, Mainz, Germany; M. Lafontaine, Institut National de Recherche et de Securité, Vandoeuvre, France; S. Ludeman, Department of Chemistry, The Catholic University of America, Washington, USA; R. Massey, Ministry of Agriculture, Fisheries and Food, Norwich, England; M. de Méo, Laboratoire de Microbiologie, Faculté de Pharmacie, Marseille, France; M. Miraglia, Instituto Superiore de la Sanità, Rome, Italy; R. Mitchum (present address), Quality Assurance Division, US Environmental Protection Agency,

Las Vegas, USA; H.C. Van der Plas, Laboratory of Organic Chemistry, Wageningen, The Netherlands; D. Reed, Department of Biochemistry and Biochemistry and Biophysics, Oregon State University, Corvallis, USA; P. Van Roosmalen, Alberta Workers' Health and Safety and Compensation, Edmonton, Alberta, Canada; E.B. Sansone, Environmental Control and Research Program, NCI-Frederick Cancer Research Facility, Frederick, MD, USA; P.L. Schuller (present address), Keuringsdienst Van Waren Door Het Gehied, Goes, The Netherlands; M.G. Siriwardana (present address), Gesira University, Wad Medani, Sudan; P.L.R. Smith, British Food Manufacturers Industries Research Association, Leatherhead, England; B. Spiegelhalder, German Cancer Research Centre, Institute of Toxicology and Chemotherapy, Heidelberg, Germany; D. Spincer, Imperial Group Ltd., Bristol, England; A. Stacchini, Instituto Superiore di Sanita, Rome, Italy; L.A. Sternson (present address), Pharmaceutical Research and Technologies, Chemical Research and Development, Philadelphia, USA; G.M. Telling, Unilever Research, Sharnbrook, England; S.P. Tucker, National Institute for Occupational Safety & Health, Robert A. Taft Laboratories, Cincinnati, USA; M. Vahl, Ministry of Environment, National Food Institute, Soborg, Denmark; J.J. Vallon, Laboratoire de Chimie Analytique Pharmaceutique, Faculté de Médecine et de Pharmacie, Lyon, France; T. Vu Duc, Institut Universitaire de Médecine, Lausanne, Switzerland; E.A. Walker, Imperial College of Science and Technology, Department of Chemistry, London, England, K. Webb, Laboratory of the Government Chemist, London, England; D.T. Williams, Health and Welfare Canada, Ottawa, Ontario, Canada, for their active participation in this programme.

CHAPTER 7

Handling Hazardous Pharmaceuticals

M.G. LEE

1 Introduction

It can be argued that all pharmaceuticals are hazardous since, at high enough doses, all drugs will exert toxic effects. But provided drugs are supplied and administered in accordance with the manufacturer's product data sheet,[1] the risks to pharmacist or patient are negligible for the majority of these substances. It is possible for some people to suffer adverse reactions at doses which are far below the therapeutic range of the drug, for example the allergic reactions that occur in persons sensitive to penicillin or phenothiazines. Provided, however, that the person involved is aware of his or her sensitivity, it should be possible to avoid the causative agent. There are, nonetheless, two classes of pharmaceuticals to which exposure should be minimized in the laboratory or hospital environment. These are cytotoxic drugs and radio-pharmaceuticals.

2 Cytotoxic Drugs

It is now well recognized that most anticancer drugs are potentially hazardous substances since they are either mutagenic, teratogenic or carcinogenic. There is evidence that patients can develop secondary tumours as a result of treatment with antineoplastic agents.[2,3,4] It must be assumed, therefore, that there is a potential threat to the health of any persons exposed to this class of drugs. Whilst such risks may be deemed acceptable for patients with life-threatening diseases, it is clearly undesirable for health care workers to be exposed to hazardous chemicals during their routine work.

One of the major problems with cytotoxic drugs is that the long-term effects of continuous exposure to very low, sub-therapeutic levels are not easily definable or fully understood. There is, however, a large body of evidence to show that health care personnel involved in the preparation and administration of anticancer preparations can, if not adequately protected, absorb potentially harmful amounts.

Much of this evidence comes from epidemiological studies on nursing and pharmaceutical staff who have handled cytotoxic agents. Studies have shown an association between the spontaneous abortions and foetal abnormalities in nurses and occupational exposure to cytotoxic drugs.[5,6] Pharmacists who were reconstituting anticancer drugs excreted mutagenic urine over the period of their exposure. When they stopped handling the drugs, the mutagenic activity declined to the level of unexposed controls within two days.[7]

Detectable amounts of cyclophosphamide have also been found in the urine of nurses preparing and administering this substance.[8] The two nurses involved were working in a cancer clinic and took no special precautions when handling the drug. Further evidence of the need for control measures is provided by two studies which found detectable amounts of fluorouracil in the air of a drug preparation room.[9,10]

It can be concluded, therefore, that pharmacy and nursing staff who prepare cytotoxic drugs for administration are potentially at risk. Consequently, it is necessary to adopt measures that will protect staff from occupational exposure. Because it is not possible to establish maximum safe exposure levels, all appropriate steps should be taken to prevent, or at least reduce to a minimum, exposure to these hazardous substances in the workplace. These include the following measures:

- Prevention or control of exposure using suitable protective cabinets.

- Staff monitoring and health surveillance by occupational health specialists.

- The use of approved written procedures.

- On-going staff training.

- Procedures for dealing with spillages and disposal of contaminated waste.*

A key factor when dispensing injections of cytotoxic drugs is the need to achieve a balance between operator protection and product protection. Asepsis and the maintenance of sterility is of paramount importance with this group of drugs, because their immuno-suppressive activity reduces the patient's ability to overcome infection. This need for sterility assurance, together with the requirement for the prevention or control of the risks associated with the handling of cytotoxic drugs, has led to the introduction of centralized pharmacy-based cytotoxic reconstitution and dispensing services within hospitals.

One of the consequences of centralization has been a re-evaluation of the control systems available for such services, and the subsequent development of more efficient and effective safety cabinets and isolators designed specifically for dispensing cytotoxic drugs. Whilst the centralization of cytotoxic preparation reduces the risks for health care workers who previously directly handled the drugs, the risk is increased for the small numbers of staff who operate the centralized facility. Adequate protection facilities and health surveillance are, therefore, important for these staff. The following discussion describes these two aspects of preparing and dispensing cytotoxic injections.

*Methods for the disposal of carcinogenic waste are reviewed in Chapter 6.

3 Safety Cabinets for Handling Cytotoxic Injections

3.1 Vertical Laminar Flow Cabinets (VLFC)

The design of VLFCs is based upon that of the class II microbiological safety cabinet.[11] A vertical down flow of laminar flow air, filtered through a High Efficiency Particulate Air (HEPA) filter (99.997% efficiency), passes over the work surface. The air then passes through vents at the front and back of the cabinet followed by a second HEPA filter, and is recirculated.

Approximately 30% of the recirculated air is exhausted from the cabinet and, to compensate for this, air is drawn in through the front opening. This creates a negative pressure within the cabinet, and the balance between the cabinet downflow and the air drawn in at the front of the cabinet produces an air curtain. The operator and product protection properties result from the presence of this air curtain. The air exhausted from the cabinet, may be recirculated into the room or ducted to the outside (Figure 1).

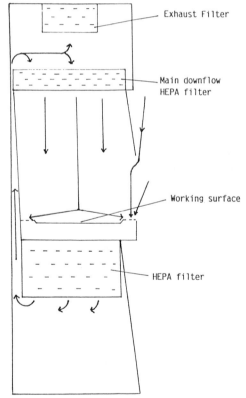

Figure 1 *Diagrammatic representation of the airflow pattern in a class IV safety cabinet*

Although the British Standard on which the design of the cabinet is based makes reference to vertical laminar flow protection cabinets, this standard is not readily applicable to hazardous drugs. Bacteria have a defined mass or bulk and are of known particle size whereas cytotoxic contaminants will be of variable size and may be present as a solid, liquid, vapour, or aerosol, and for materials handled in microbiological safety cabinets, operator protection is more critical than product protection.

The siting of VLFCs is important. Any air turbulence at the front of the cabinet will disturb the air curtain and thus reduce the operator and product protection factors for the cabinet. Such turbulence may be caused by air draughts from ventilation systems, opening and closing of doors in the room in which the cabinet is situated, or by personnel walking past the front of the cabinet. Correct operating procedures are, therefore, essential for the efficient operation of this type of cabinet.

Despite the few disadvantages of VLFCs, there is clear evidence of their effectiveness in controlling exposure of employees to cytotoxic agents. Pharmacy staff handling cytotoxic drugs in purpose-built cytotoxic units showed no increase in urinary mutagenicity compared with unexposed controls.[12] Similarly, the mutagenic activity in the urine taken from staff working in oncology units and pharmacies was significantly reduced after the provision of adequate safety precautions.[13]

Using more sophisticated monitoring techniques, it has been demonstrated that pharmacists working in a fully-protected environment do not absorb drugs, although on one or two occasions there has been some evidence of exposure.[14] More recently, it has been confirmed that pharmacy staff preparing cytotoxic injections in a VLFC with 30% recirculation and wearing gloves and arm protection, had no increase in urinary mutagenic activity compared with controls.[15] Abnormal haematological values have been found, however, in pharmacists who formulated cytotoxic drugs in a faulty biohazard cabinet.[16] Low neutrophil counts (and chromosomal aberrations) were indicative of increased exposure to a cytotoxic agent.

A further consideration with VLFCs is that, for optimum sterility assurance they should be sited in a clean room conforming to BS 5295 class J.[17] This in turn means the operator should wear sterile clean room clothing. Such requirements can result in higher costs for a centralized facility, and have led to the application of 'total barrier techniques' using isolator technology.

3.2 Isolators

Isolators are totally enclosed work stations supplied with HEPA filtered air (99.997% efficiency). Operators use glove ports or a half-suit arrangement to access the working area. Materials are introduced either through a transfer chamber or by using an access port and docking device. During operation the cabinet is totally sealed from the outside. Isolators can have a rigid or flexible structure and their design can have a considerable impact upon their potential uses.

For safe cytotoxic reconstitutions the pressure inside the isolator must be lower than the external environment, and for this reason isolators for use with cytotoxics generally have a rigid construction.

The main advantage of isolators is that they can be located in an unclassified environment. If flexible film isolators are used in these locations, however, care should be taken to prevent pinholes in the PVC film and the resulting leakage of contaminated air into the isolator.

Two types of rigid isolator currently available for cytotoxic reconstitution and dispensing are shown in Figures 2 and 3. Some variation in design is possible to allow for different working practices, but the basic requirement is that the pressure in the working enclosure should be lower than that of the transfer chamber and the external environment. Provided all seals and surfaces are intact, the greatest risk to operator and product protection occurs during the transfer or materials into and out of the working enclosure. For this reason, careful thought needs to be given to the design of the transfer system, and the transfer procedure should be carefully validated. For a detailed review of the design and operation of isolators see Goulding et al.[18]

Isolators offer an economic and safe alternative to VLFCs for centralized cytotoxic reconstitution services. They can be housed in an unclassified environment and a minimum of protective clothing is required. The efficiency is not affected by air turbulence around the cabinet since the work area is totally enclosed. For safety reasons, they should be located within a designated room or in a designated area within the pharmacy.

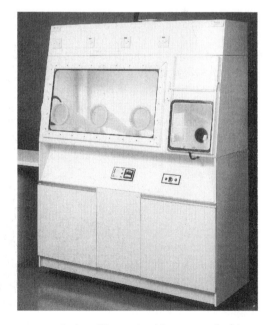

Figure 2 *Compact cytotoxic handling suite (Amercare Ltd.)*

Figure 3 *Container cytotoxic dispensing cabinet (Envair (UK))*

4 Staff Monitoring

A system of health surveillance is necessary for all staff directly involved with the routine handling of cytotoxic drugs. The Control of Substances Hazardous to Health (COSHH) Regulations (1988) require *inter alia* that, for defined areas of risk, appropriate health surveillance procedures should be introduced, and where necessary exposure should be monitored. For staff involved in the reconstitution, preparation, and administration of cytotoxic agents, surveillance should include regular general health screening in conjunction with specific biochemical and cytogenetic tests.

4.1 Health Surveillance

Health surveillance is essential for staff directly involved in handling cytotoxic drugs. The COSHH Regulations require employers to introduce appropriate health surveillance and, where necessary, monitor staff exposure to hazardous substances. There are clear risks associated with cytotoxic drugs which need to be controlled and monitored. Staff monitoring should consist of regular general health screening together with biochemical or cytogenetic tests to determine if an individual has been exposed to harmful levels of mutagenic substances.

Fundamental to any occupational health monitoring programme is an interview for all new staff working with cytotoxic drugs together with a physical examination.

The interview is intended to review the employee's medical history but can also be used as an opportunity for the member of staff to discuss personal worries concerning their work with potentially hazardous substances. Simple laboratory tests, for example differential white cell counts, urine and liver function tests, may also be performed to confirm the employee's current state of health. This interview should be repeated at suitable intervals; annual interviews are recommended.

The concept of an individual exposure record has been proposed as an attempt to quantify the drug 'burden' to which an individual has been exposed. Such data are, however, difficult to collect accurately, since the drugs vary in their dosage and acute toxicity and it would not be possible to differentiate between exposure to different drugs. A record of the time spent by individuals handling cytotoxic agents is simpler to produce and probably more relevant than a cumulative record of the doses an individual has prepared at work. These exposure records should also include specific details of any involvement in accidental spillages or any other incident which could have enhanced exposure. In addition, all such incidents should be reported to the occupational health specialist.

Physical examination should be a routine pre-employment practice, and should be repeated at regular intervals in order to identify symptoms associated with acute exposure, for example, irritation of mucous membranes, dizziness, or light-headedness. Similarly, following accidental spillage or exposure, a physical examination should be undertaken. Documentation of untoward events together with the exposure record should form a part of individual health surveillance records. Records should be maintained centrally, ideally by occupational health or personnel departments. In addition, exposure and incident records should form part of the employee's personnel record.

Blood and differential white cell counts are performed routinely in many centres. Unfortunately, data from such tests give little useful information except in cases of high exposure, and must be interpreted with care. Analysis of specific drugs or metabolites in body fluids is possible and has been reported for specific agents. The example quoted earlier of cyclophosphamide detected in the urine of two nurses handling the drug without special precautions[8] is one such case. The same investigation also demonstrated that cyclophosphamide can be absorbed through intact skin.

The presence of thioethers in urine has been used as an indicator of exposure to alkylating agents. The mean urinary concentration of thioethers was higher in a group of oncology nurses following a five-day work rotation than it was when they returned to work after a three-day leave.[19] The value of such tests is probably limited, due to the increasing range of therapeutic cytotoxic agents available and the low levels expected *in vivo* from occupational exposure. It was not possible, for example, to detect platinum in the urine of pharmacists or nurses who frequently prepared or administered cisplatin.[20] But this may indicate the success of the safety precautions employed when handling that drug.

4.2 Cytogenetic Monitoring

Cytogenetic tests can potentially offer a sensitive indicator for assessing the true biological risks to staff of exposure to cytotoxic drugs, but the test methods currently available appear to be either insufficiently sensitive or poorly validated. Tests which measure mutagenic activity or identify cytogenetic changes have been used to monitor staff exposure, and details are given below. For a more detailed assessment of their use in the context of hospital staff exposure to cytotoxic drugs see Ferguson *et al.*[14] and Kaijser *et al.*[21]

4.2.1 *Test for Urine Mutagenicity*[15]

The Ames test is used routinely to measure mutagenic activity in urine. The test is based upon the detection and measurement of mutations in bacterial cells *in vitro* caused by agents in the urine. The test suffers, however, from a lack of both sensitivity and specificity, since there are many agents, other than cytotoxic drugs, which increase the mutagenic activity of urine. Smokers, for example, show increased urinary mutagenicity. Consequently, the test is now considered unsuitable for monitoring staff.

4.2.2 *Assay for Chromosome Aberrations*[14]

This method offers the most direct estimate of cytogenetic changes. It is, however, low in sensitivity, laborious to perform, and requires a high level of technical skill. The test measures long-lasting effects and is capable, therefore, of giving an index of long-term damage. If a test does reveal an increase in chromosomal aberrations with time for a particular individual, this probably indicates that hazardous substances are being absorbed, and urgent preventative measures are required.

4.2.3 *Micronucleus Assay*[14]

Micronucleus assays are less time consuming than chromosomal aberration tests and do not require the same high level of technical expertise. The assay is a more indirect method of detecting exposure, which, due to the variability, will not identify differences in individuals. It may, however, identify differences between populations or working conditions, if applied to sufficiently large groups of workers.

4.2.4 *Sister Chromatid Exchange (SCE)*[14]

This method detects reciprocal exchanges between chromatids. It can detect changes caused by very low levels of mutagens and is more sensitive than other cytogenetic techniques. Since SCE lesions are short lived and decline within a few days, the test must be undertaken immediately or within a short time of sample collection. The test has no retrospective value and is applicable to staff whilst they are working with cytotoxic drugs.

4.3 Summary of Cytogenetic Tests Used for Monitoring

It is not clear which, if any, of the cytogenetic tests should be used for staff monitoring. The SCE test would appear to be the most relevant for personnel who prepare cytotoxic preparations. If a general screen of all staff, irrespective of current duties, is required then the chromosomal aberration test is the only appropriate method.

Cytogenetic monitoring of human populations is both expensive and time consuming. Reproducibility of data and test interpretation can cause problems. Numerous biological and environmental factors may induce chromosomal changes and confound the data.

The major limitation to the inclusion of cytogenetic measurements in routine medical surveillance is how to interpret individual results. It is not possible to interpret accurately a positive urine mutagenicity test or cytogenetic result in prognostic terms. Also, it is not possible to link definitely a single positive result to occupational exposure. A positive result may provoke unjustifiable anxiety in individuals for whom its importance cannot be adequately explained. It can be argued, however, that a monitoring programme would be one method of determining whether there is a relationship between the consistent detection of abnormalities and occupational exposure to cytotoxic drugs. Cytogenetic tests may also provide a warning of equipment failure, poor technique, or inadequacies in protective clothing. The most realistic application of cytogenetic tests may well prove to be as a quality assurance tool to identify overall levels of exposure.

5 Radiopharmaceuticals

Assessment of the risks associated with handling radiopharmaceuticals is less complicated than for cytotoxic agents. The hazards associated with ionizing radiation are well documented and realistic occupational exposure limits can be applied. Monitoring techniques are also adequately validated. Continuous monitoring using film badges and thermoluminescent dosimetry provides an accurate record of the exposure of staff engaged directly in radiation work.

The pharmaceutical considerations that apply to cytotoxic agents and the need for asepsis are of equal importance with radiopharmaceutical preparations. The protective cabinets available for use with cytotoxic preparations have been developed from cabinets that were originally designed for radiopharmaceutical applications. Much of the foregoing arguments are, therefore, equally applicable to the facilities for handling the radiopharmaceuticals.

The disposal of radiopharmaceuticals is controlled under the Radioactive Substances Act and the Control of Pollution (Radioactive Waste) Regulations 1976. These specify the amounts of radioactive waste that hospitals may dispose of over a given time together with the method of disposal.

For a fuller discussion of the hazards associated with handling radiopharmaceuticals see Lee.[22]

6 References

1. 'ABPI Datasheet Compendium', Datapharm Publications Ltd., London 1993.

2. S.M. Seiber and R.H. Adamson, *Adv. Cancer Res.*, 1975, **22**, 57.

3. J.D. Boice, M.H. Green, J.Y. Killen, S.E. Ellenberg, R.J. Keehn, E. McFadden, and T.T. Chen, *New Eng. J. Med.*, 1983, **309**, 1079.

4. P.D. Berk, J.D. Goldberg, M.N. Silverstein, A. Weinfield, P.B. Donovan, J.T. Ellis, S.A. Landaw, J. Laszlo, Y. Najean, A.Y. Pisciotta, and L.R. Wasserman, *New Eng. J. Med.*, 1981, **304**, 441.

5. S.G. Selevan, M.L. Lindbohm, R.W. Hornung, and K. Hemminki, *New Eng. J. Med.*, 1985, **313**, 1173.

6. K. Hemminki, P. Kyronom, and M.L. Lindbohm, *et al.*, *J. Epidemiol. Comm. Health*, 1985, **39**, 141.

7. C. Macek, *J. Am. Med. Assoc.*, 1982, **242**, 11.

8. M. Hirst, S. Tse, D.G. Mills, and L. Levin, *Lancet*, 1984, (i), 186.

9. A. Neal, R.A. Wadden, and W.L. Chiou, *Am. J. Hosp. Pharm.*, 1983, **40**, 597.

10. M.L. Kleinberg and M.J. Quinn, *Am. J. Hosp. Pharm.*, 1981, **38**, 1301.

11. British Standard 5726 Specification for Microbiological Safety Cabinets, British Standards Institution, London, 1993.

12. J. Cooke, *Pharm. J.*, 1987, **239**, R2.

13. B. Kolmodin-Hedman, P. Hartvig, and M. Sorsa, *Arch. Toxicol.*, 1983, **54**, 25.

14. L.R. Ferguson, R. Everts, M.A. Robbie, V. Harvey, D. Kempel, D. Mak, and A.J. Gerred, *Aust. J. Hosp. Pharm.*, 1988, **18**, 228.

15. E.P. Guinee, G.H. Beuman, G. Hageman, I.J. Welle, and J.C.S. Kleinjans, *Pharm. Weekblad. Sci. Ed.*, 1991, **13**, 72.

16. P. Rodriguez and C.Y. Yap, *Aust. J. Hosp. Pharm.*, 1991, **21**, 39 (letter).

17. British Standard 5295 Environmental Cleanliness in Enclosed Spaces, Parts 0, 1, 2, 3, British Standards Institution, London, 1989.

18. N.M. Goulding, M.G. Lee, and A.C. Moore, in 'The Cytotoxics Handbook', 2nd Edition, ed. M. Allwood, and P. Wright, Radcliffe Medical Press Ltd., Oxford, 1993, p. 9.

19. O. Jagun, M. Ryan, and H.A. Waldron, *Lancet*, 1982, (ii), 443.

20. S. Venitt, C. Crofton-Sleigh, J. Hunt, *et al.*, *Lancet*, 1984, (i), 74.

21. G.P. Kaijser, W.J.H. Underberg, and J.H. Beijnen, *Pharm. Weekblad. Sci. Ed.*, 1990, **12**, 212.

22. M.G. Lee, in 'Risk Assessment of Chemicals in the Environment', ed. M.L. Richardson, The Royal Society of Chemistry, London, 1988, p. 491.

CHAPTER 8

The Handling and Disposal of Radioactive Materials

B. KERSHAW

1 Introduction

The handling of radioactive materials is a prerequisite for much current biomedical and other research. Work with radioactive substances in a laboratory environment is perhaps more stringently controlled than some aspects of chemical research, and this is reflected in the UK legislation for the handling and disposal of radioactive materials. This chapter is intended to serve as an *aide-mémoire* for managers and Radiation Protection Supervisors who have to oversee experiments involving radioactive materials.

1.1 Radioactive Materials

The Ionising Radiations Regulations (IRR) 1985[1] impose requirements on all holders and users of radioactive materials for the identification, checking, and recording of such materials. Radioactive materials can be classified into two groups — unsealed dispersable radioactive material and sealed sources. The following definitions apply:

A *radioactive substance* is any substance which has or is suspected to have an activity concentration greater than 100 Bq g^{-1}.[1]

A *sealed source* is a radioactive material wholly bonded within a solid inactive material or encapsulated in a receptacle so that no leakage can occur during storage or foreseeable conditions of use.[1]

Unsealed dispersable radioactive material is radioactive material not considered to be a sealed source.

1.2 Accounting for Radioactive Sources

Accounting procedures need not be adopted for any radioactive material having a half-life of less than 3 hours *or* in the form of contamination *or* where the amount of a discrete source or confined radioactive material has an activity less than the value specified in Column 2, Schedule 2 of the Ionising Radiations Regulations 1985.[1]

Data in Table 1 show the level of activity for a selection of radionuclides *at* or *above* which accounting procedures are required for the use of that material. Table 1 also contains some physical properties of these radionuclides which relate to their measurement and safe-handling, and the annual limits on intake by inhalation. The IRR 1985 should be consulted for other radionuclides.

Table 1 *Minimum activity levels for selected radionuclides at or above which accounting procedures are required, some selected physical properties, and annual limits on intake by inhalation*

Atomic number	Radionuclide	Minimum quantity for accounting (Bq)[a]	Specific activity (Bq/g)	Half-life	Annual limit on intake by inhalation (Bq)
1	Tritium	5×10^6	3.59×10^{14}	12.262 y	3×10^9
6	Carbon-14	5×10^5	1.70×10^{11}	5568 y	9×10^7
15	Phosphorus-32	5×10^5	1.07×10^{16}	14.3 d	1×10^7
15	Phosphorus-33	5×10^5	5.99×10^{15}	24.4 d	1×10^8
16	Sulfur-35	5×10^6	1.58×10^{15}	87.1 d	8×10^7
17	Chlorine-36	5×10^5	1.19×10^9	3.08×10^5 y	9×10^6
27	Cobalt-57	5×10^5	3.15×10^{14}	270 d	2×10^7
27	Cobalt-58	5×10^5	1.16×10^{15}	72 d	3×10^7
27	Cobalt-60	5×10^4	4.22×10^{13}	5.24 y	1×10^6
34	Selenium-75	5×10^5	5.33×10^{14}	121 d	2×10^7
43	Technetium-99	5×10^6	6.33×10^8	2.12×10^5 y	2×10^7
53	Iodine-125	5×10^4	6.44×10^{14}	60 d	2×10^6
53	Iodine-129	5×10^6	5.99×10^6	1.72×10^7 y	3×10^5
53	Iodine-131	5×10^4	4.55×10^{15}	8.14 d	2×10^6
55	Caesium-137	5×10^5	3.63×10^{12}	26.6 y	6×10^6
80	Mercury-197	5×10^5	9.07×10^{15}	65 h	3×10^8
80	Mercury-203	5×10^5	5.07×10^{14}	46.9 d	3×10^7
88	Radium-226	5×10^3	3.63×10^{10}	1622 y	2×10^4
90	Natural thorium	5×10^3	8.14×10^3	1.39×10^{10} y[b]	40[b]
92	Uranium-238	5×10^6	1.23×10^4	4.5×10^9 y	2×10^3
94	Plutonium-239	5×10^3	2.26×10^9	2.4×10^4 y	200
95	Americium-241	5×10^3	1.20×10^{11}	458 y	200

[a]Taken from Column 2, Schedule 2 of The Ionising Radiations Regulations 1985, Statutory Instrument 1333: 1985, HMSO, London, 1985.
[b]Thorium-232

Records for accounting for any particular radioactive substance should contain:

- A means of identification unique to that source;

- Date of receipt;

- Activity at a specified date;

- Location — updated at appropriate intervals;

- Date and manner of disposal when appropriate.

An annual check should be carried out to ensure that the accounting record is a true one.[2]

1.3 Storage of Radioactive Material

Each laboratory or area must have a recognized defined storage area for radioactive materials in use. This area must be provided with protection from mechanical and fire hazards as well as providing adequate shielding and containment.

The nature of the radioactive material being stored, in any type of containment, must not generate a hazard as a function of storage time, *e.g.* pressurization due to a build-up of gas from radiolytic decomposition, thermal problems *etc*. Some radioactive decay products are more hazardous than the parent material, in which case the properties of the decay products will govern the storage mode.

Solutions of radioactive materials also undergo radiolytic decomposition, which may lead to a dangerous increase of pressure in a sealed container. Glass, silica, and some plastic containers can become seriously embrittled by the radiation which together with any pressure increase or induced chemical reaction could cause them to break catastrophically. Solutions of radioactive materials should therefore be stored in appropriate radiation resistant material containers which can be readily and regularly vented, be given some form of secondary containment, have no stress points such as graduation or volume indication marks etc, and not contain chemically unstable or reactable mixtures.

1.4 Leak Testing of Sealed Sources

All sealed sources must be leak tested at least once every 26 months or more frequently dependent upon the conditions and usage to which the source is subject.[1] The requirements for leak testing are detailed in the Approved Code of Practice IRR 1985 Part 1.[2] The test must conform to the British Standard BS 5288:1976, Appendix D.

2   Methods of Protection

Every attempt must be made to limit the degree of exposure achieved through working with radioactive materials or radiation sources. Table 2 lists the types of radiation which may be encountered.

Their range in air varies with their nature and energy and gives rise to two types of radiological hazard — *internal* and *external.*

Table 2 *Main types of ionising radiations and their hazards*

Radiation type	Main hazard to personnel	Protection
Alpha particles	Internal	Containment
Beta particles	Internal and external skin dose	Containment, local shielding and exposure time
Gamma and X-rays	External	Distances, shielding and exposure time
Neutrons	External	Special shielding and exposure time
Charged particles or ions	External	Shielding and exposure time

2.1 The External Radiation Hazard

2.1.1 Basic Principles of Control

The basic methods of protection against external radiation are:

- Restriction of the strength of every source to the minimum necessary for the task in hand;

- The use of the maximum amount of distance between the source and the operator, compatible with the satisfactory and safe performance of the work;

- Restriction of the period of exposure to the minimum compatible with safe working;

- The use of suitable shielding.

The protection necessary in any particular situation to ensure that doses are kept below the relevant limit may be achieved by a combination of these methods.

2.1.2 The Use of Distance

The intensity of radiation from a radioactive source decreases with increasing distance. For a point source (and where the dimensions of the source are small compared with the distance from the source to the point of interest), the dose rate is inversely proportional to the square of the distance, *i.e.* by doubling the distance the dose rate is reduced by a factor of 4 and so on. For example, the gamma dose rate from a 1 GBq cobalt-60 source decreases with increasing distance as follows:[3]

at   1 cm   ≈   3.5 Sv h⁻¹
at 10 cm   ≈   35 mSv h⁻¹
at 100 cm   ≈   350 µSv h⁻¹

Radioactive sources should therefore never be handled with bare hands, or with gloved hands unless the thickness of the glove is sufficient to reduce the radiation to reasonable levels.

2.1.3 The Use of Time

A simple calculation

$$\frac{\text{acceptable dose} \quad \text{(Sv)}}{\text{dose rate per hour} \quad (\text{Sv h}^{-1})}$$

gives the working time in hours.

The acceptable dose must be kept both within statutory dose limits (IRR 1985)[1] and ALARP — As Low As Reasonably Practicable.

Exposure to high dose rates calls for careful pre-planning, and sometimes for 'dummy' runs. 'On the job' discussions in a radiation field should be avoided.

2.1.4 The Use of Shielding

Beta Radiation. The most suitable shielding materials for beta radiation are sheets of light metals such as aluminium or Perspex. The absorption of beta particles in matter gives rise to *bremsstrahlung* radiation (electromagnetic radiation resulting from the retardation of charged particles). For sources of energetic beta radiation, a combination of perspex and lead makes the best shielding material.

Gamma rays cannot be completely absorbed by a shield; they are only reduced in intensity. Of course, any degree of attenuation is possible if the shield is made thick enough. The approximate thicknesses of various materials required to attenuate 1 MeV gamma rays by a factor of 10 are:[4]

Lead	**Iron**	**Concrete**	**Water**
3.5 cm	6 cm	20 cm	40 cm

(These figures refer to broad beam geometry)

Thermal neutrons are completely absorbed by cadmium. This absorption process is accompanied by the emission of gamma rays and it may be necessary to back up the cadmium with lead. Boron-containing material (*e.g.* Boral) is also a good absorber of thermal neutrons.

Fast neutrons. The dose rate from a given flux of thermal neutrons is much less than the dose rate from the same flux of fast neutrons. The first step therefore in the shielding of fast neutrons is to reduce the energy until they can be completely absorbed as indicated above for thermal neutrons. This is best accomplished by using materials made up of light elements, *e.g.* hydrogenous materials such as paraffin wax, wood and concrete.

2.2 The Internal Radiation Hazard

An internal radiation hazard arises when radioactive material is taken into the body through the nose or mouth or through breaks in the skin. Radionuclides behave in the body in the same way as the corresponding stable element. For example, radiocalcium will be deposited in bone by the same pathway as stable calcium; tritium behaves like hydrogen and becomes incorporated in water and organic constituents.

Behaviour in the body will depend further on the chemical form of the material taken in, its particle size and the route of entry.

Following entry a number of organs may be irradiated until the material is eliminated from the body or until the radioactivity decays. Because the range of alpha particles in tissue is far smaller than that of beta particles or gamma rays, their energy is deposited in a smaller volume close to the site of deposition and the biological damage caused is correspondingly greater. For this reason, and because of their general long half life, alpha emitting materials in the body are most hazardous. This is reflected in the protection methods necessary to contain them.

2.2.1 Routes of Entry

Ingestion. Contamination on surfaces may lead to ingestion of activity through the mouth. Control is based on a combination of rules and procedures and strict laboratory discipline, *e.g.* in the correct use and removal of gloves, correct monitoring procedures after working in contaminated areas, and no eating, smoking, drinking, taking snuff, or applying make-up in contamination-controlled areas.

Inhalation. Work carried out in a laboratory or workshop can be accompanied by the formation of airborne contamination. The assessment of the significance of radioactive airborne contamination is a difficult problem due to the influence of many factors such as breathing characteristics (rate of breathing, whether the individual breathes through the nose or the mouth *etc.*), the size, shape and density and the chemical properties of the airborne particles (which will affect lung deposition and subsequent metabolism), and the ventilation pattern in the working area. Control is largely based on proper containment and ventilation coupled with correct working discipline. Before a job is carried out consideration must be given to the possibility of airborne contamination.

Absorption. Radioactive contamination may penetrate the skin by diffusion through the skin barrier or *via* cuts and wounds. Diffusion applies particularly to tritiated water (both as a vapour and as a liquid). Radioactive materials deposited on the skin and absorbed through the skin may subsequently disperse *via* the blood stream. Organic solvents are particularly dangerous in that they can penetrate the skin easily. In general, however, the skin forms an efficient barrier to contamination.

Cuts and wounds may allow active materials direct access into the bloodstream. If caused in potentially contaminated areas they should be washed in warm running water with plenty of soap and mild bleeding should be stimulated. Decontamination of healing skin injuries is very difficult.

Control is based largely on correct laboratory discipline and techniques, *e.g.* when using solvents *suitable* gloves should be worn.

2.2.2 Basic Principles of Control

There are three main principles of control against the internal radiation hazard:

• Containment;

• Cleanliness;

- Use of the least toxic radioactive material that is suitable, and the minimum quantities in all experiments.

  Containment. When containment is lost, radioactive material can be rapidly dispersed with the consequent risk of it being taken into the body. Work likely to give rise to radioactive contamination (*e.g.* handling of powdered radioactive materials) should therefore be carried out under conditions affording adequate containment.

The two methods of containment of operations most widely used are partial containment by means of *fume cupboards,* and complete containment by means of *glove boxes.*

2.3 Working in Fume Cupboards

In a radiochemical laboratory, fume cupboards are frequently used for work with low level radioactivity as well as for more conventional chemical work.

Fume cupboards present an intermediate level of containment, for work with radioactive materials, between fully enclosed glove box or shielded cell facilities and open laboratory bench operations. They are also a useful facility for opening packaged items which have been previously cleaned to a low but often unknown contamination level and removed from active glove boxes.

Before any operation is carried out in a fume cupboard check that:

- The quantity of radioactivity being handled is suitable for this containment, *i.e.* confirm a glove box or shielded cell is not required;

- There are no items present which may prove hazardous, *e.g.* sharp instruments;

- Air is flowing into the fume cupboard. A paper tissue held at the face gives a rough check. The flow across the fume cupboard face should be between 0.7 — 1 m/sec;[5]

- There is adequate lighting;

- Correct personal and environmental monitoring are available. Set up contamination monitoring equipment close to the fume cupboard in such a way that gloves can be checked without touching anything outside the cupboard.

The following *working practices and procedures* should be adopted for fume cupboard work:

i)   As far as possible ensure that the areas in front of fume cupboards are kept clear of all obstructions;

ii)   The user should *never* allow his/her head to enter a fume cupboard;

iii)   Rubber gloves must always be worn for operations within a fume cupboard. Check that laboratory coat cuffs are tucked into gloves. After any fume cupboard work and prior to touching any item outside the fume cupboard, gloves should be either discarded in the fume cupboard or monitored, taking care not to contaminate the monitors. Any glove found to be contaminated should be left in the fume cupboard and disposed of in due course as radioactive waste. Gloves must not be kept in laboratory coat pockets;

iv)   Take care not to spread contamination from the fume cupboard. Any item to be removed from a radioactive fume cupboard must be directly monitored where possible and/or smear-monitored to ensure that it is free from any loose contamination before it is allowed to come into contact with any surface outside the fume cupboard;

v)   Drip trays having sufficient volume in the event of spillage should be used when handling liquids. The fume cupboard floor should also be lined with absorbent paper;

vi)   You must have receptacles in the fume cupboard correctly marked for aqueous liquid waste, organic liquid waste, and solid waste. If you have to use 'sharps' or have any broken glass, then these must be disposed of into a clearly labelled 'sharps' container, ideally an aluminium screw-capped can;

vii)   When opening packages of items posted from an active glove box it is essential to adopt a technique for opening the package in stages, to establish the level of activity on the surface of the item and minimize the spread of any loose contamination;

viii)   Radioactive gases released during fume cupboard work must be removed before discharging carrier or purge gases into the ventilation system wherever this is possible;

ix)   Use suitable containers for storage of radioactive and hazardous materials. Materials should be stored only for short periods of time;

x)   Radiation levels can be reduced by use of appropriate shielding. Care must be taken not to exceed the floor loading of the fume cupboard or to interfere with air flow patterns. Similar precautions should be taken with any equipment. Keep the fume cupboard tidy;

xi)   Do not move or handle unshielded sources directly by hand. Use tongs or forceps;

xii) Gloves which have been used in a fume cupboard should not be used for work outside the fume cupboard. If any contamination is found or suspected on gloves, these must not be used but removed correctly and disposed of safely.

xiii) Work must be planned such that there is sufficient time to make the fume cupboard safe, tidy up, dispose of waste, and carry out personal contamination monitoring. Monitor the sill and floor area outside the fume cupboard for alpha and beta contamination by direct and/or smear-monitoring as appropriate.

2.4 Working in Glove Boxes

The need for complete containment (*e.g.* by means of glove boxes) should be considered if it is necessary to handle more than 3×10^7 Bq of plutonium (or material of equivalent toxicity) in solution or 3×10^5 Bq as a solid. Powders should always be handled in a glove box. The type of work or the physical nature of the radioactive material may modify these figures by a factor of 10 or more either way. Advice on this sort of problem should be obtained from the Radiation Protection Adviser.

Glove boxes are often used to contain large quantities of highly toxic materials, and failure of the containment, *e.g.* by means of a damaged glove, may release some of this toxic material and lead to airborne contamination perhaps hundreds of times the Derived Air Concentration. It is therefore necessary for glove box operators to give their whole attention to the operations being carried out in order to avoid untoward incidents and for them to examine their box gloves at frequent intervals for signs of failure. Glove box operators should wear surgical gloves (in addition to the box gloves) and these surgical gloves should be checked for contamination each time the operator finishes a period of work in the box.

The following *working practices and procedures* should be adopted for glove box work:

i) In general these will be along the same lines as those for working in fume cupboards. However, it is important to remember that the effectiveness of the glove box as a means of protection relies heavily on complete containment. As such introducing or removing material and equipment from the box must be carried out *via* sealed systems *e.g.* bag posting using heat sealers;

ii) It is important to carry out frequent personal contamination monitoring and to monitor the external surfaces of all packages bagged out of a box thoroughly;

iii) It may occasionally be necessary to wear additional protective equipment, *e.g.* respirators, for some posting operations.

2.5 General Procedures for Work with Radioactive Materials

The following *working practices and procedures* should be adopted:

i) The area, whether it is a laboratory, a fume cupboard, or the inside of a glove box, should be maintained in a tidy and orderly state;

ii) There should be no unnecessary accumulation of radioactive materials;

iii) Benches and floors should be covered with PVC sheet or bitumenized paper when there is a risk of spilling radioactive material;

iv) Any surface contamination arising during an operation should be cleaned-up immediately;

v) Sources surplus to requirements should be disposed of or returned. Waste should be disposed of in a recognized safe way.

2.6 Personal Protective Equipment

The IRR 1985[1] require that in an area which is supervised or controlled due to contamination hazards, personnel should be issued with appropriate personal protective equipment (PPE). The minimum requirements will be dictated by the contamination hazards which exist.

If you intend to cross a contamination control barrier to enter a contamination controlled area you should:

• Leave your 'clean side' shoes on the clean side and put on your personal 'active side' shoes or put on overshoes;

• Once on the active side you must also put on an active area laboratory coat or coveralls;

Immediately prior to leaving the active side of the barrier you must:

• Monitor your laboratory coat or coverall before removing it, and if clean leave them on the active side; monitor any other item to be taken across the barrier;

• Monitor your shoes and leave them on the active side of the barrier;

• If you are using overshoes these must be monitored prior to removal at the barrier and placed in appropriate waste bin.

2.7 Special Protective Equipment

The IRR 1985[1] require that in areas of high levels of airborne contamination, respiratory protective equipment (RPE) should be available. This equipment must be of a type and standard approved by the Health and Safety Executive (HSE).

All RPE must be thoroughly examined at suitable intervals (not generally to exceed 1 month) and properly maintained. A suitable record of this examination must be kept for at least 2 years.

RPE will include respirators, airhoods and pressurized suits. Advice must be sought from the Radiation Protection Adviser on the most appropriate equipment for the operation.

3 Medical Surveillance

The following procedures relating to medical surveillance should be adopted:

- Before any person can become a classified radiation worker (see Section 5.1) he or she must undergo a medical examination to ensure that he or she is fit to commence such work;

- Once classified, periodic reviews must be undertaken (at least once every 12 months) to ensure that he or she remains fit;

- Facilities need to be provided to an Employment Medical Adviser (EMA) or Appointed Doctor (AD) to allow medical surveillance to be carried out;

- Any employee who receives an over-exposure in excess of twice any annual dose limit should undergo a special medical examination. The EMA or AD must be notified of any over-exposure — *i.e.* one greater than the dose limit;

- A health record must be kept for at least 50 years from the date of the last entry made in it;

- An employee must co-operate with medical surveillance requirements and must provide such information as may be reasonably required.

4 Monitoring for Radiation and Contamination

Monitoring for radiation and contamination falls into three categories:

i) *Personal monitors* which, as their name implies, are carried on the person and hence give a measurement of the radiation or air contamination to which the person is exposed;

ii) *Portable monitors*, usually battery operated, can be moved from place to place as the need arises. They are used in particular for carrying out detailed measurements at various positions during specific operations, and also for carrying out routine surveys;

iii) *Installed monitors*, which are at fixed positions, are usually mains operated and are used to monitor the *general* radiation and air contamination levels in the working environment, or personal contamination.

4.1 Monitoring in the Workplace

Suitable monitoring equipment must be obtained *prior* to starting work and must be available during the work period. If there is any doubt regarding the equipment to be used the Radiation Protection Adviser must be consulted.

The following points should be borne in mind:

* During active handling operations (*e.g.* fume cupboards or glove boxes) monitoring should be carried out frequently as work proceeds, and at the end of a working session prior to leaving the laboratory;

* Gloved hands and laboratory coat sleeves and other parts which might become contaminated must be monitored;

* Ungloved hands, laboratory coats and feet should be monitored before leaving the workplace and proceeding to the contamination-control barrier;

* All shoes, hands and clothing must be monitored when leaving a contamination-controlled area whether or not active work was performed. Appropriate alpha and beta radiation monitoring instruments must be made available in the laboratories and at the barriers.

4.2 Monitoring of the Workplace

The IRR 1985[1] require monitoring of Radiologically Designated Areas (RDAs) to be carried out. Monitoring is one important means of indicating whether levels of radiation or contamination are satisfactory for continuing work with ionizing radiation, detecting breakdowns in control of systems, and detecting changes in levels of radiation. Thus monitoring needs to take place outside controlled and supervised areas as well as inside in order to check that such places remain correctly designated.

A schedule to monitor the workplace for levels of radiation and contamination relevant to the designation of the area, and arrangements for implementing it, should be set up.

4.3 Personal Radiation Monitoring (External Exposures)

4.3.1 The Film Badge

The film badge is sensitive to beta, gamma, X- and slow neutron radiation. The response of the unfiltered photographic emulsion to gamma radiation is energy dependent and it is largely unaffected by neutrons. Filters in the film badge holder make it possible to determine doses due to radiations of the following types and energies both singly and in mixtures:

- X- and gamma radiation;

- Beta radiation;

- Slow neutrons.

4.3.2 Thermoluminescent Dosimeters (TLDs)

Thermoluminescent materials such as lithium fluoride (LiF) release light when they are heated after exposure to beta or gamma radiation and can therefore be used for the measurement of dose. TLDs are particularly useful when measuring extremity dose, or when rapid results are required.

4.3.3 Neutron Monitoring Dosimeters

There are several types of dosimeter available for measuring neutron dose. The most appropriate dosimeter needs to be selected depending on dose and energy. CR39*, track films, and neptunium fission methods are available.

4.3.4 Personal Alarm Monitors

A variety of pocket-sized personal gamma integrating dose and dose-rate alarm monitors are available. These devices may be fitted with a digital display, an audible alarm, a wide range of sensitivity options, energy filtered detectors, and have non-threshold and threshold modes of operation.

---

*CR39 is a type of plastic. Exposure to neutrons produces 'pits' in CR39, which are electrochemically etched and counted to assess the dose.

4.4 Personal Internal Contamination Monitoring (Internal Exposures)

4.4.1 Personal Air Samplers

These monitors consist of a small battery operated air pump (enclosed in a case) which can be placed in a laboratory coat or coverall pocket or held in a harness. Attached to the pump is an air sample head which contains a glass fibre filter paper. The position of the air sample head is arranged so that exposures as near as possible to the breathing zone may be measured. Air is drawn through the filter paper (normally at about 2 litres per minute). After use the amount of radioactivity deposited on the filter paper is measured in a counter of known efficiency.

4.4.2 Personal Biological Monitoring (Bioassay)

Internal dose assessments can be made by examining excreta from the body. The methods available involve monitoring nose blows, urine and faeces.

Any person who, through his routine daily work, may possibly receive an intake of radioactive material may be asked to provide samples routinely (*e.g.* urine, faecal) for analysis. The frequency and type of analysis are decided upon by consideration of the materials that people could come into contact with and in what quantities.

4.4.3 Whole Body Monitoring

Where applicable *in vivo* measurements can be carried out using a body monitor. This relies on photons of sufficient energy being present inside the body to penetrate the body tissues and be detected outside the body.

4.5 Portable Radiation Monitoring Equipment

These instruments use ionization chambers, Geiger-Müller counters, proportional counters or scintillation counters as the detector according to the type and range of the ionizing radiation to be measured. A wide selection of battery operated monitors are available for the measurement of dose rates in the range from natural background of about 0.1 μSv h^{-1} to 50 Sv h^{-1}. Energy response, monitor range, and sensitivity to ionizing radiations are important factors which should be taken into account before selecting and using a particular monitor in an ionizing radiation environment.

Portable neutron monitors which employ a thermal neutron proportional counter filled with a Helium-3 gas mixture and surrounded by a large mass of moderating material are usually used for measuring neutron dose equivalent rates of up to 10 mSv h^{-1} over the energy range thermal to 11 MeV.

4.6 Portable Surface Contamination Monitoring Equipment

These instruments use a Geiger-Müller counter, scintillation counter, or proportional counter detector coupled to a suitable counting ratemeter.

The more common alpha detectors consist of a zinc sulfide screen in which scintillations are produced by the alpha particles. A photomultiplier tube converts the scintillations into electrical pulses of suitable magnitude for operating the ratemeter. The zinc sulfide screen is sensitive to light, and the screen assembly is protected by a thin aluminized membrane (1 mg cm^{-2}) which excludes light but permits entry of alpha particles. Detection efficiencies of greater than 15% are normal with these detectors.

Alpha-sensitive semiconductor probes consisting of silicon surface barriers with sensitive areas of 0.5 cm^2 to 44 cm^2 are also available for alpha surface contamination monitoring.

Beta scintillation detectors which combine good detection efficiency together with a large detection area are increasingly replacing Geiger-Müller counters for beta surface contamination monitoring.

Monitoring of contaminated surfaces may be achieved by direct means or by smear techniques. Smear-monitoring is carried out if it is not practicable to place the probe in contact with the surface, if the contamination levels are too high, or if the background radiation is too high. The estimation of surface contamination levels by smear-monitoring is complicated by having to make an assumption as to the fraction removed from the surface; this is usually assumed to be 10%. It is essential to resort to smear-monitoring for alpha activity on wet surfaces and the smears must be dry before being measured.

4.7 Portable Air Contamination Monitoring Equipment

These instruments contain a pumping unit which draws air through a particulate collecting medium (normally a glass fibre filter paper) or directly into an ionization chamber. In the former type the sampling medium on which the air contaminants have been collected is returned to the laboratory for measurement. For sampling in a worker's breathing zone the sampling head can be attached to an anglepoise arm.

Gaseous Monitors are suitable for the measurement of the concentration of inert fission product gases and tritium. These instruments which are battery or mains operated incorporate a filter in order to exclude particulate contamination from the ionization chamber detector.

4.8 Installed Radiation Monitoring Equipment

Installed Radiation Monitors are used to measure gamma, X- or neutron radiation dose rate in working environments where radiation fields are likely to change rapidly as a result of non-routine (or in some cases routine) work operations or incidents. In special cases they can be used for obtaining general environmental monitoring information around new plants and at the boundaries of or outside Radiologically Designated Areas. A variety of monitors with Geiger-Müller or ionization chamber detectors are available for the measurement of dose rate in the range from natural background of about 0.1 μSv h^{-1} to 10^4 Sv h^{-1}.

4.9 Installed Personnel Contamination Monitoring Equipment

Installed Personnel Contamination Monitors are designed for measuring alpha and/or beta-gamma contamination on the hands, shoes and clothing of workers. The hand and clothing monitor is an instrument into which the worker can insert his hands and which monitors simultaneously, *via* a pair of dual phosphor scintillation counters (above and below each hand), the level of alpha and beta contamination. Separate units are available with alpha or beta detector probes for monitoring clothing.

4.10 Installed Air Contamination Monitoring Equipment

Installed Air Contamination Monitors are used to measure airborne contamination levels at fixed positions in the working environment. These monitors are usually mains operated and contain a pumping unit.

Particulate Dust Samplers draw air through a particulate collecting medium, normally a glass fibre filter paper (charcoal paper for radioactive iodines), which is returned to the laboratory for measurement and assessment.

Particulate Alarm Monitors monitor airborne contamination *in situ*. Separate monitoring systems are used for measuring alpha and beta-gamma active dust. These monitors have an alarm setting which can be adjusted according to the monitor's sensitivity to the Derived Air Concentration (DAC) in Bq m^{-3} of the contaminant being monitored.

Gaseous Monitors which monitor and record the concentration of inert fission product gases and tritium *in situ*.

5 Statutory Regulations

5.1 The Ionising Radiations Regulations 1985

The Ionising Radiations Regulations[1] were laid before Parliament during September 1985 and came into operation on the 1 January 1986. A supporting Code of Practice was approved by the Health and Safety Commission at about the same time.

The Regulations are the most comprehensive legislation on radiological protection ever produced in the United Kingdom and they apply to everyone who works with sources of ionizing radiations.

Some of the main requirements of the regulations are as follows:

i)   An employer shall take all necessary steps to restrict as far as reasonable practicable the exposure of persons to ionizing radiation. This is the principle of ALARP which has the same objective as ALARA (as low as reasonably achievable) used by the International Commission of Radiological Protection (ICRP);[6]

ii) An employer shall ensure that his employees and other persons are not exposed to radiation in excess of the dose limits summarized in Table 3;

iii) Radiation Protection Advisers must be appointed to advise on radiological safety matters and the requirements of the Regulations;

iv) Radiologically Designated Areas must be identified where there is a likelihood of persons receiving doses in excess of three-tenths (Controlled Area) and one-tenth (Supervised Area) of the annual dose limit for workers aged 18 years or over;

v)   All persons who enter or work in Controlled Areas must be designated *'Classified'* or must be operating under a *written system of work* agreed with the Radiation Protection Adviser;

vi) Classified persons (and others) will be subject to dose monitoring and appropriate medical surveillance. Dosimetry services have to be approved by the Health and Safety Executive;

vii) An employer is required to keep certain records, *e.g.* radiation dose records, health records, records of quantities and location of radioactive substances, records of investigations into exposure and incidents etc;

viii) Employers must ensure that persons working with ionizing radiation have received adequate information, instruction and training for the safe conduct of that work in accordance with the Regulations;

ix) Radiation protection arrangements must be covered in written local rules, and work involving ionizing radiation must be supervised. This last requirement necessitates the appointment of *Radiation Protection Supervisors;*

x)   Any equipment, device or other article used in connection with ionizing radiation must afford optimum radiation protection; designers and manufacturers etc. have a duty to ensure that their products can be used in a way so as to reduce doses ALARP;

xi) Every employer must monitor levels of ionizing radiation using properly maintained and tested (once every 14 months) equipment. Tests must be carried out by a Qualified Person (QP).

Table 3 *Statutory dose limits for occupational exposure*

Age	Wholebody mSv y^{-1}	Individual organs and tissues mSv y^{-1}	Lens (eye) mSv y^{-1}
18 or over	50	500	150
Trainees under 18	15	150	45
Any other person	5	50	15
Woman of reproductive capacity	—	*Abdomen* Not pregnant: 13 mSv in any consecutive 3-monthly interval Pregnant: 10 mSv throughout declared term of pregnancy	—

Notes to Table 3:

a) No person under the age of 18 can be designated a classified worker.
b) Whenever an individual's whole body dose exceeds 15 mSv in a calendar year an in-house investigation must be carried out into that individual's working conditions, the object of which is to demonstrate that the dose was as low as reasonably practicable. The investigation report must be made available to the HSE if requested.
c) The Regulations allow for exposure in excess of the above Statutory Limits in emergency situations.

5.2 Transport Regulations

The International Atomic Energy Agency (IAEA) Regulations for the Safe Transport of Radioactive Materials[7] form the basis for UK legislation governing the movement of radioactive materials.

The Secretary of State for Transport at the Department of Transport, London, is the appointed UK Competent Authority, and the Regulations are administered through a Radioactive Materials Transport Division, headed by the Transport Radiological Advisers.

A 'Code of Practice for the Carriage of Radioactive Materials by Road'[8] has been prepared by the Department of Transport to assist all concerned, and consignors in particular, to discharge their obligations under the law.

6   Disposal of Radioactive Waste

Radioactive waste is defined as any unwanted material contaminated by, or containing radioactive materials, or unwanted radioactive materials themselves.

The production and accumulation of radioactive waste must be minimized and recorded. It must be categorized and assessed prior to safe storage. It is essential to control the accumulation, processing and storage of radioactive wastes to ensure their suitability for eventual disposal and to restrict their discharge to within legally authorized limits.

Three types of radioactive waste need to be considered:

* Gaseous/particulate in air discharges to the atmosphere;

* Liquid discharges; and

* Solid waste.

6.1 Discharges to Atmosphere

Under the Radioactive Substances Act (1960)[9] no radioactive waste should be discharged to atmosphere other than in accordance with an authorization granted by Her Majesty's Inspectorate of Pollution (HMIP) and, where necessary, the Ministry of Agriculture Fisheries and Food (MAFF). The authorization gives an activity limit in a specific period. All discharges must be kept ALARP using best practical means (BPM). This places an obligation on the operator to ensure that he continually reviews his arrangements for controlling discharges and ensuring adequate dispersal.

The methods available are:

i)   Removal of particulate activity by filtration;

ii) Reduction of gaseous activity by scrubbing or adsorption;

iii) Dispersal *via* a high stack.

Adequacy of controls can be assessed by means of environmental monitoring programs.

6.2 Liquid Discharges

Where the level of radioactivity is low, liquid waste can be discharged to sewers or directly into a river. This must be done with the regulatory consent of HMIP and MAFF. Like discharges to atmosphere, an authorization to discharge must be granted under the Radioactive Substances Act (1960)[9] and BPM must once again be used to control discharges.

The methods available are:

- Treatment by ion-exchange, evaporation, or chemistry to concentrate the radioactivity into a residue suitable for conversion to solid waste;

- Allow decay of shorter lived radionuclides.

Adequacy of controls can be assessed by means of environmental monitoring programs.

6.3 Solid Radioactive Waste

Radioactive waste production must be minimized and any waste should be assigned to the lowest possible category. Segregation of isotopes should be carried out wherever possible in order to facilitate waste streaming. The 'Guide to the Administration of the Act'[10] associated with the Radioactive Substances Act 1960 gives details of disposal routes.

6.3.1 Very Low Level Wastes

Solid wastes <0.4 Bq g^{-1} can be exempted from consideration under the Act. In this case a Low Activity Substances Exemption Order will be made.[10]

6.3.2 Trench Burial

Land burial has been practised for some years at the Drigg site, south of Sellafield. The wastes authorized to be disposed of at Drigg are subject to many restrictions, details of which can be found in the 'Guidance to the Administration of the Act'.[10]

6.3.3 National Disposal Service

Some wastes are not suitable for local disposal. Arrangements have been made by the UK Department of the Environment for dealing with these wastes. Wastes can be collected by, or sent to, the National Disposal Service at Harwell.

6.3.4 Domestic Refuse

Small amounts of solid waste are authorized with ordinary refuse. Alpha emitters and Sr^{90} are usually excluded from 'dustbin' disposals.

The limits for such disposal with ordinary refuse are: 400 kBq in any 0.1 m^3 and 40 kBq per article.[10]

6.3.5 Incineration

This can be used for combustible, obnoxious or biologically toxic radioactive wastes. Authorizations for disposal in this way are given taking all local conditions into account.

7 References

1. The Ionising Radiations Regulations 1985, Statutory Instrument 1333:1985, HMSO, London, 1985.

2. Approved Code of Practice 'The Ionising Radiations Regulations 1985; The Protection of Persons Against Ionising Radiation Arising from any Work Activity', HMSO, London, 1985.

3. 'Handbook of Radiological Protection', Prepared by a Panel of the Radioactive Substances Advisory Committee, Pt 1:DATA, HMSO, London, 1971.

4. 'Recommendations for Data on Shielding from Ionising Radiation, Pt 1: Shielding from γ Radiation', BS 4094 Pt 1, British Standards Institution, London, 1966.

5. A.E.C.P. 1054 'Ventilation of Radioactive Areas; Standards Section', AEA, Risley, April, 1989.

6. 'Recommendations of the International Commission on Radiological Protection', *ICRP Publications 26*, Ann. ICRP, 1, No. 3, Pergamon Press, Oxford, 1981.

7. International Atomic Energy Agency, Safety Series No. 6, 'Regulations for the Safe Transport of Radioactive Material', Revised Edition, Vienna, 1985.

8. Department of Transport, 'Code of Practice for the Carriage of Radioactive Materials by Road', HMSO, London, 1978.

9. Radioactive Substances Act 1960, HMSO, London, 1960.

10. 'Radioactive Substances Act 1960; A Guide to the Administration of the Act', HMSO, London, 1960.

Acknowledgement

Thanks to Mr. A. Mills and Dr. J.A. Berry for their valuable wisdom and experience incorporated in this chapter.

CHAPTER 9

The Handling and Disposal of Infectious Waste

C.H. COLLINS

1 Introduction

Infectious waste is discarded material that contains, or may contain pathogens — microorganisms that are capable of causing disease. Most infectious waste is generated in clinical and biomedical laboratories that investigate material from patients who are suffering from microbial diseases. Some pathogens are present in the environment, however, albeit in small numbers, and may be cultured inadvertently in laboratories that are engaged in, for example, quality control of raw materials, feeds and foodstuffs. This applies in particular to laboratories that use commonplace culture media within the pH range 6.5 — 7.8 and incubation temperatures between 30° and 39° C. A single colony of, *e.g.* a salmonella, may contain enough organisms to constitute several infectious doses.

2 Nature of Infectious Laboratory Waste

Laboratory waste from clinical laboratories is included among the various official definitions and categories of 'clinical waste', *i.e.* waste that arises during healthcare activities.[1,2] Unfortunately these are not detailed, and differing types of infectious laboratory waste are placed in different categories. A list of laboratory waste[3] has been published, however, and that part which includes infectious waste is shown in Table 1. It cannot be claimed that this is inclusive but it will provide a basis for risk assessment.

3 Legal Aspects and Risk Assessment

Pathogenic microorganisms are 'substances hazardous to health' under the 'Control of Substances Hazardous to Health Regulations 1988'[4] and are 'biological agents' according to the EC 'Directive on the Protection of Workers from Exposure to Biological Agents at Work'.[5] These two pieces of legislation will be harmonized in the near future. They require employers to identify hazardous substances, to assess the risks they pose to workers and others, and to take the necessary steps to reduce, minimize or eliminate those risks.

Table 1 *Infectious waste generated by microbiological laboratories*[a]

- Specimens or their remains, in their containers, containing blood, faeces, urine, secretions, exudates, transudates, other normal or morbid fluids, samples of food in food.

- All cultures made from these specimens, directly or indirectly.

- All other cultures made from foods, feeding stuffs and raw materials that have been incubated at 37° C in media at pH 6.5 — 8.0.

- All stocks of microorganisms that are no longer required.

- Used diagnostic kits.

- Used disposable transfer loops and rods, pasteur pipettes, and slides.

- Disposable cuvettes and containers used in chemical analysis of morbid or pathological materials.

- Biologicals standards and microbiological control materials.

- Paper towels and tissues used to wipe hands, equipment and benches.

- Disposable gloves, gowns and aprons.

- Disposable sharps; hypodermic needles and syringes, scalpels, blades, broken glass, ampoules.

- Tissues and animal carcasses.

- Bedding and refuse from animal cages.

---

[a]Adapted from Collins and Kennedy[3]

3.1 Risk Assessment

No laboratory has complete control over the microbial content of the samples or specimens that it receives for examination. Microorganisms have been categorized, however, according to the hazards they offer to those who work with them and to the general public. The categorization systems adopted by different countries vary in their wording but are the same in principle: four groups, numbered 1 — 4, the lowest

group being relatively harmless and the highest the most hazardous. Table 2 shows
the system formulated by the UK Advisory Committee on Dangerous Pathogens
(ACDP).[6] Lists of organisms in Groups 2 — 3 are published[6] and are revised from
time to time. An organism not listed in these groups is considered to be in Group 1.
It is expected that the UK and other European lists will be harmonized in the near
future. Very few laboratories handle pathogens in Group 4, and then only under the
supervision of the ACDP and the Health and Safety Executive, and are therefore not
considered here.

Table 2 *Classification of microorganisms on the basis of hazard*[a]

Group	Description
1	Unlikely to cause human disease.
2	May cause human disease; may be a hazard to laboratory workers but exposure rarely causes infections; unlikely to spread in the community; treatment and prophylaxis available.
3	May cause serious human disease and be a hazard to laboratory workers; may spread in the community; treatment and prophylaxis available.
4	Causes severe human disease; offers serious threat to laboratory workers; high risk of spread in the community; no effective treatment or prophylaxis.

[a]Adapted from the Advisory Committee on Dangerous Pathogens;[6] organisms in
Classes 2, 3, and 4 are considered to be pathogens.

    In this chapter we are concerned with these organisms only when they are
included in waste. For risk assessments of organisms during handling and
experimental investigations, readers are invited to consult other works.[7,8]

3.1.1 Organisms in Group 1

Although these organisms are most unlikely to cause human disease, they should not
be ignored. Exposure to waste that contains large numbers of them may be hazardous.
For example, some fungi in this group are used to evaluate the resistance to spoilage
of manufactured goods such as fabrics. Some of these organisms release large

numbers of spores. In addition, the inhalation of spores or metabolites of some Group 1 bacteria may induce allergies.

3.1.2 Organisms in Group 2

Although many of the organisms in this group are incriminated in human disease they are also naturally present in the environment, *e.g.* in soil, water, foodstuffs and indeed in any kitchen. Although samples of those materials offer a minimal risk, any assessment must take into account the numbers of cultures containing organisms recovered from them.

3.1.3 Organisms in Group 3

Waste containing these organisms, whether specimens or cultures, are hazardous and careful assessment is required.

3.1.4 Organisms and Parts of Organisms Used in Genetic Manipulations

Most of these organisms are either in or from bacteria in Group 1 and do not therefore qualify as infectious. Nevertheless, as genetic manipulation is an emotive subject, it is prudent to treat waste containing them in the same way as other microbiological waste.

4 Minimization and Elimination of Hazards

Minimization of hazards is required while the waste is being generated. Elimination of hazards is necessary after the investigation is complete and the material is discarded.

4.1 Minimization

The objective is to 'contain' the waste materials and protect laboratory and other staff from contact with the hazard. This is achieved by chemical disinfection; the agents used in most microbiological laboratories are phenolics, hypochlorites, aldehydes, and alcohols. Choice depends on the nature of the organisms and care must be taken to use the correct disinfectant for the purpose; not all disinfectants are effective against all microorganisms, either in their vegetative state or as spores.[9]

Small articles, such as slides, cover glasses, disposable loops, and pipettes are placed in the disinfectant pots and jars which are part of the normal equipment of a laboratory work bench. Any article put into disinfectant must be completely submerged, and air bubbles must be removed so that the fluid is in contact with all surfaces.

Hypodermic needles and other 'sharps' should be consigned to the standard sharps discard boxes as provided in the National Health Service.[10]

Larger objects, *e.g.* petri dishes and culture tubes, should be collected in autoclavable discard bags or bins. Autoclavable bags are usually transparent and colour-coded with blue-print wording. They should be supported in shallow bins (made of autoclavable plastic) in case they are overloaded and burst.[9] Objects that are too large to be put into these bags may be placed directly in the bins.

At the end of each working day (or more frequently in busy laboratories), the disinfectant containers, bags and bins should be taken to the preparation room for final treatment and disposal. The necks of the bags should be closed with 'quick ties' and the bins covered to prevent spillage during transit.

4.2 Elimination — Making the Waste Safe

'It is a cardinal rule that no infected material shall leave the laboratory.'[11]

Since this caveat first appeared in print in 1974 there has been a general agreement among microbiologists that infectious waste from laboratories should never enter the normal municipal solid waste stream. The ACDP[6] makes a distinction between waste material containing Group 1 organisms and that containing Groups 2 and 3 pathogens. Group 1 material that is not destined for incineration should be 'rendered non-viable' but that in Groups 2 and 3 'must be 'made safe' before disposal or removal to the incinerator. The Health Services Advisory Committee,[13] however, although endorsing the autoclaving of waste acknowledges that 'if it is not reasonably practical to autoclave it' the waste can be transported directly to an incinerator, even presumably if that incinerator is some distance away and transport may be on the public highway. It is difficult to conceive of a microbiological laboratory that does not have an autoclave, but the hazards of transporting infectious laboratory waste are considerable.

There are two options then for making the waste safe: autoclaving and incineration.

4.2.1 Autoclaves

Autoclaves are essential items of equipment in microbiological laboratories and ideally each laboratory should have two, one for sterilizing culture media and equipment and the other for sterilizing infectious waste. The waste pots, bags and bins should be collected from the laboratories and all ties and covers removed before they are placed in the autoclave; otherwise steam may not penetrate the load.

The usual laboratory autoclave cycle for decontamination is 121 °C for 15-20 min holding time at temperature[6,9] but this may be varied under biological control. Much higher temperatures are required for the destruction of prions and unconventional agents.[6]

4.2.2 Incinerators

Infectious waste should be incinerated directly only if the incinerator is under the immediate control of the laboratory staff.[9,11]

Incineration processes are under continual review but the current official guidance[12] is a temperature of 850-900 °C in the primary chamber and not less than 1000 °C with a gas residence time of not less than 2 seconds in the secondary chamber. Incinerator emissions must comply with the complex legal requirements under the control of pollution and environmental protection legislation.

5 Organization and Final Disposal

5.1 Organization

The laboratory organization should ensure that infectious waste and materials for reuse are effectively separated and that there is no possibility of infectious material being returned to the laboratory.

Figure 1 is a flow chart of safe practices for treating infectious laboratory waste, and Figure 2 proposes how these may be translated into laboratory design.[9] Graduated glass pipettes require exceptional treatment, as the only items of equipment for which chemical disinfection will suffice. They should remain in contact with the disinfectant for at least 18 hours and then be washed with hot water (60° C).

5.2 Final Disposal

Figure 1 suggests that some of the waste could be macerated and flushed to the public sewer *via* a plumbed-in waste disposal unit. Before this is contemplated, however, it is necessary to consult the local office of the Rivers Authority, which may place limits on the quantities and nature of waste disposed of in this way.

Autoclaved waste is safe and in theory could enter the municipal solid waste stream which is usually landfilled. This would be unwise, however, as some of the waste could still be identifiable and the general public, not knowing that it is safe, might be concerned. It is best to place the autoclaved waste in the yellow bags[1] used for clinical (hospital) waste and destined for incineration, and arrange for that method of final disposal.

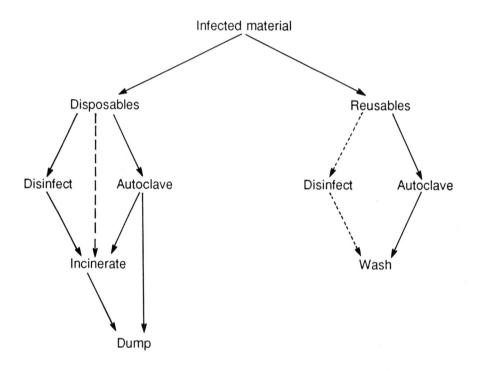

———— Normal practice

— — — Incinerator under laboratory control

------- Graduated pipettes

Figure 1 *Flow chart for the treatment of infected material*
(Reproduced with permission from C.H. Collins, 'Laboratory Acquired Infections',
3rd edition, Butterworth-Heinemann, Oxford, 1993).

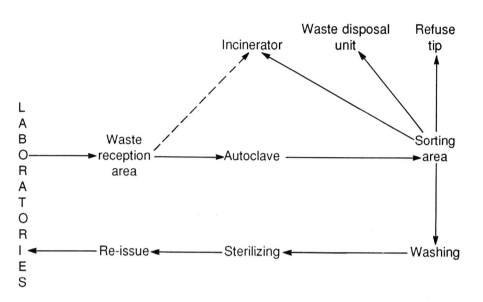

- - - - Only if incinerator is under laboratory control

Figure 2 *Design of preparation (utility) rooms. Flow chart for the disposal of infected laboratory waste and reusable materials*
(Reproduced with permission from C.H. Collins, 'Laboratory Acquired Infections', 3rd edition, Butterworth-Heinemann, Oxford, 1993).

6 References

1. Health Services Advisory Committee, 'The Safe Disposal of Clinical Waste', HMSO, London 1992.

2. Department of the Environment, 'Clinical Wastes'. Waste Management Paper No. 25, HMSO, London, 1983.

3. C.H. Collins, and D.A. Kennedy, 'The Treatment and Disposal of Clinical Waste', Science Reviews, Leeds, 1993.

4. 'The Control of Substances Hazardous to Health Regulations 1988, Approved Code of Practice,' 2nd Edition, HMSO, London, 1990; Statutory Instrument 1657:1988.

5. European Directive, 'The Protection of Workers from Risks Related to Exposure to Biological Agents at Work. 90/679/EEC', Brussels.

6. Advisory Committee on Dangerous Pathogens, 'Categorization of Pathogens According to Hazard and Categories of Containment', 2nd Edition, HMSO, London, 1990.

7. C.H. Collins, *Letters in Appl. Microbiol.*, 1990, **10**, 109.

8. C.H. Collins, in 'The COSHH Regulations: a Practical Guide', ed. D. Simpson, and W.G. Simpson, The Royal Society of Chemistry, Cambridge, 1991, p. 95.

9. C.H. Collins, 'Laboratory Acquired Infections', 3rd Edition, Butterworth-Heinemann, Oxford, 1993.

10. 'Specification for Containers for Sharps', Department of Health, London, 1982.

11. C.H. Collins, E.G. Hartley, and R. Pilsworth, 'The Prevention of Laboratory Acquired Infections', Public Health Laboratory Service Monograph No. 6, HMSO, London, 1974.

12. Secretary of State's Guidance on Clinical Waste Incinerator Processes under 1 tonne per hour. PG5/1, Department of the Environment, London, 1991.

13. Health Services Advisory Committee, 'Safe Working and the Prevention of Infection in Clinical Laboratories', HMSO, London, 1991.

CHAPTER 10

Laboratory Waste — A Wasting Asset

M.P. SATCHELL

> '*Recycling is often held to be a good practice in itself. The truth,
> however, is that its value — like that of any activity — should be
> decided only following a proper assessment*'[1]

1 Introduction

With the increase in public concern about the environment, one area where attention
is particularly being focused is the impact of waste material, both from the point of
view of the effect of its disposal, and the environmental implications for the
production of replacement material. Furthermore, all raw materials absorbed by a
shop, factory, or laboratory, are part of that organization's resources, and in these
days of financial stringency it is necessary to make optimum use of these assets.
Waste products have to be disposed of at minimum cost to the organization, with the
least harmful impact to the environment, and in such a way as to comply with current
environmental legislation.

Waste minimization and *recycling* are two ways to achieve these aims — the one
attempting to reduce the amount of waste produced, the other to make the optimum
use of it when it does arise. Waste minimization is very much a matter of local
conditions involving the working practices, processes or procedures employed in any
particular area of the office, factory, or laboratory, and can best be implemented by
those closest to the situation.

Waste recovery and recycling involve wider areas of interest and rely on the
premise that one man's waste is another man's commodity. An excellent overview of
the main forms of waste recycling currently in use is available.[1]

In order to demonstrate how recycling can work in practice, some of the schemes
introduced in my own factory site at Merck Ltd. in Poole, Dorset, will be described.
This is a medium sized chemical factory on the south coast, away from the main
industrial conurbations and includes analytical and R & D laboratories. The main
offices are also within a mile of the site.

2 Waste Recovery and Waste Recycling

2.1 Audits

The conventional first step in any waste minimization is to perform a 'Waste Audit'. When I first initiated a waste audit at Poole, waste was put into three large skips which were emptied three times a week (at £13 per skip) and subsequently went for landfill. By establishing what actually went into these skips and determining what recycling systems were already in place within the company, I was able to set up a series of schemes to divert a major part of the waste from the refuse tip into usable and often profitable channels within the company.

2.2 Recycled Materials

2.2.1 Metal Containers

25 l metal solvent drums had previously been thrown away. But if they had only contained clean solvent, these drums could easily be collected together, cleaned, and reused. Drums that could not be reused were crushed and, provided they were free from any hazardous contents, sold as scrap metal. Open-top drums and 200 l drums are used where possible for storing waste chemicals before being sent to a chemical disposal contractor.

Many chemicals are sent to us in aluminium cans, and some of Merck's products are stored in aluminium tubs, which are reused several times. These containers are eventually disposed of as scrap aluminium along with a variety of other aluminium items such as small pots for laboratory chemicals, furnace tubes from the Inorganic Development laboratory, and discarded fire extinguishers (cut in half for safety). We try to ensure that the aluminium is free from other metal components particularly iron. The presence of other metals would not prevent recycling but would reduce the price paid for scrap aluminium. We also collect empty aluminium drinks cans from both sites.

Stainless steel and copper/brass are also collected in separate skips and by so doing we again ensure the best price for scrap. If we were to consign all these different metals to a single skip we would only obtain the scrap value of mild steel (~£15 per tonne), whereas stainless steel and aluminium separately can currently command £240—300 per tonne while the price per tonne for copper scrap is ~£1000. This demonstrates one of the essential points on recycling practice — segregation at source. The 'cleaner' any item, the easier it is to recycle and therefore the higher its scrap value. By keeping different types of waste separate the recycler does not have the problem and cost of sorting and separation (and disposal of contaminants) so the economics of recycling is more attractive.

2.2.2 Glass

Glass from the factory and the laboratories is a mixture of colours and often contaminated with chemicals. Until recently we had been able to dispose of all our waste glass free of charge by sending it (along with our waste chemicals) to a nearby disposal company who use the waste glass to reline their incineration furnaces. We now send our (segregated) clean brown and white glass for recycling. We hope soon to be able to reuse our 2.5 l Winchester bottles for refilling with solvents and acids.

2.2.3 Plastic

Some plastics are collected for recycling, and again segregation is essential. Loose polystyrene packing material used for goods delivered to us by other suppliers is saved and reused for our own packaging. In fact, apart from periods of high output, the site is self-sufficient in second-hand loose polystyrene. 'Bubblewrap' wrapping is also saved and reused.

2.2.4 Computer Paper and Cardboard

We use many tonnes of computer paper in a year and this paper is collected and sent for recycling. Paper suitable for recycling should be free from paper-clips, 'stickies', 'Sellotape'®, and carbon paper. Office paper can command a good price if segregation from extraneous items is properly carried out. Some waste paper merchants will remove unsegregated waste paper free of charge, provided certain items are excluded, for example, carbon paper, envelopes, magazines or books with glue binding, and FAX paper. White and coloured paper should also be segregated.

Our own cardboard boxes are reused several times before disposal, and almost all waste cardboard is then sent for recycling. Waste paper merchants do make a small charge for collection, but the cost is less than if paper and cardboard are sent with the other waste to a landfill site.

2.2.5 Printer and Photocopier Accessories

A rapidly expanding area for recycling is the recovery of toner cartridges from laser printers. Many companies will accept empty cartridges, and clean and refill them for less than the cost of new cartridges. Alternatively, some companies will pay about £2—3 for empty cartridges. The quality of reconditioned toner cartridges is similar to new ones. Other companies offer a free disposal service for printer cartridges and some photocopier cartridges, both of which if disposed of in the conventional way could introduce harmful chemicals into the environment at the landfill sites. (It is possible that legislation will be introduced shortly banning the landfill disposal of such cartridges.) Ribbon printers and typewriter ribbons can be reinked as a contract service.

2.2.6 'Toxic' Metals

Mettler-Toledo provide an environmentally acceptable service for the recovery of silver from discarded pH and redox electrodes. £1 is donated to the Royal National Lifeboat Institution for every electrode received.

Waste mercury has little scrap value (50p per kilo) but regularly appears from all around the site, particularly from spillages and broken thermometers. After recovery and cleaning, the mercury can be reused in vacuum gauges.

Ion-exchange resin is used in a number of our manufacturing processes, but after successive regenerations the resin has disintegrated to such an extent that it has to be discarded. It has, however, been shown to be suitable for treating some of our aqueous waste and spent resin is now used routinely for the removal of heavy metals from the liquors from several processes rendering the aqueous waste sufficiently pure to be disposed of as normal effluent, and thus enabling us to meet the stringent requirements of the National Rivers Authority.

Another area that we are considering seriously is 'computer scrap'. Discarded computers and many other electronic units may contain significant quantities of precious metals and although only low prices are likely to be paid for the whole units the separated circuit boards can be of surprising value.

2.2.7 Solvent Recovery

Recycling of solvents reduces the volume of waste that has to be sent for external disposal, and can be carried out in-house, or for large volumes of solvent by chemical contractors. Several companies will remove waste chlorinated solvents (sometimes a payment is offered), and commercial solvent recovery is available for many other common laboratory solvents.

If the solvent is to be recovered in-house, it is obviously important to ensure that the recovered material is of a quality suitable for reuse. Recovery will normally be by distillation through a fractionating column and, as a rough guide, in order to ensure the complete removal of dissolved solids, a column of 2—3 theoretical plates is required. Apart from very specific wastes, the most common use for recycled solvent in a laboratory is for cleaning apparatus.

3 Conclusion

Nationally, new recycling initiatives are continually being introduced, and the associated industries are developing organizations to collect the waste for use as a secondary raw material. In most instances the collection will be arranged through scrap merchants or waste paper companies who will advise on what is worth collecting and viable quantities for collection. Many companies will provide containers for a particular commodity.

One important piece of recent legislation that affects the transfer of all waste whether it is for disposal or recycling is the Environmental Protection Act (1990). The main features of the act, in the context of laboratory waste, are that the waste should be packed in a secure container, so that there is no risk of the contents escaping in transit, the transfer should be documented with a clear description of the nature of the contents, its source and the quantity, and that the waste should be transported by a registered carrier to a licensed site. Some carriers provide their own documentation, but this is not always the case.

As a result of the schemes described in this chapter, the volume of non-chemical waste sent out for disposal by landfill from our factory has been reduced over a period of two years from the original nine skip loads per week to two, with concomitant savings of several thousand pounds derived from reduced disposal costs, savings in packaging material purchases, plus income from the sale of scrap. The reduced impact on the environment must also be counted a benefit.

While the scale of such potential savings will vary from one organization to another, there is almost no situation where some savings cannot be made provided there is the will and the determination. In general, there is a willingness to recycle if the means and the necessary information are accessible. But beware — the enthusiasm for recycling — once aroused — can make it difficult to say 'no'.

4 Reference

1. S.M. Ogilvie, 'A Review of: The Environmental Impact of Recycling', Recycling Advisory Unit, Warren Spring Laboratory, 1992.

CHAPTER 11

Runaway Reactions

M. HANNANT

1 Introduction

When carrying out a chemical reaction or handling a chemical substance, it is the obligation of the researcher to make sure that the health and safety aspects of any procedure are known and steps are taken to minimize the risk for that procedure. By taking these precautions the researcher discharges the obligation he or she has to themselves and to their colleagues. One particular problem in the risk assessment of 'unknown' reactions is predicting the possibility of a runaway reaction occurring, and retrieving suitable information from the literature on this thermodynamic class of reaction.

This chapter sets out to help in this assessment in three ways: first by defining the term runaway reaction and looking at the main factors causing or contributing to runaway reactions; secondly by briefly discussing how to assess the risks associated with chemical reactions in general and the precautions required when carrying out a reaction; and finally by describing information sources on runaway reactions and reactive chemical hazards.

2 What is a Runaway Reaction?

All chemical reactions involve changes in energy. Most occur with release of energy (an exothermic reaction), but some take in energy on reaction (an endothermic reaction). When a reaction releases energy in a quantity too great to be dissipated by the immediate environment of the reaction system, the reaction becomes uncontrolled.

Endothermic reactions can lead to products that are energy-rich and are thermodynamically unstable. No energy is required for the products to decompose to their elements. This decomposition could also result in the release of energy.

In either case the potential for uncontrolled release of energy from an exothermic reaction or the instability associated with some endothermic products can cause a runaway reaction.

3 What are the Factors Causing or Contributing to Runaway Reactions?

There are two main factors which can contribute to a runaway reaction: the concentration of the reactants, and the temperature of the reaction.

The law of mass action states that the concentration of each reactant in a reaction mixture directly influences the velocity of the reaction and its rate of heat release. Therefore, in general, the more concentrated the reactants are, the faster the reaction proceeds. Most text books recommend the use of 10% solutions of reactants if possible. Lower concentrations of 5% or even 2% should be used for vigorous reactants and reagents, and even lower concentrations should be used when a reaction is being carried out using a catalyst. An example of an uncontrolled catalysed reaction is the violent reaction which occurred during the catalytic hydrogenation of 400 g *o*-nitroanisole at 34 bar under the excessively vigorous conditions of 250°C, 12% catalyst and no solvent. The resulting runaway reaction caused rupture of the hydrogenation autoclave.[1]

Reaction temperature influences reaction velocity, and according to the Arrhenius equation,

$\ln k = \ln A - (E_a/RT)$ where k = rate constant, A = pre-exponential factor, E_a = activation energy, R = gas constant, T = temperature,

the rate of a reaction increases exponentially with increase in temperature. A simple example of this is the second-order decomposition of acetaldehyde. Measurements between 700 K and 1000 K show that the rate of reaction doubles between 700 K and 716 K and again doubles between 900 K and 926 K.[2] For some reactions the reaction rate can double with a 10 °C increase in temperature.

An example of temperature causing a runaway reaction is the reported incident during the storage of *m*-nitrobenzenesulfonic acid. Storage at 150°C under virtually adiabatic conditions for several hours led to exothermic decomposition and a violent explosion. It was later found that the exothermic decomposition of the solution set in at 145°C and the pure acid decomposed at approximately 200 °C.[3]

One area not covered by the above factors is the tendency for one particular chemical to be more reactive than another. The instability or unusual reactivity in individual compounds can be explained in part by certain inherent structural features which are particularly unstable.

Some of these types of compounds and the particular structural features are given below.

Acetylenes	C⦂C
Alkyl or acyl nitrates	C-O-NO$_2$
Alkyl or acyl nitrites	C-O-NO
Arenediazo sulfides	C-N=N-S
Arenediazoates	C-N=N-O
Azides	N$_3$
Azo compounds	C-N=N-C
Bis(arenediazo)oxides	C-N=N-O
Bis(arenediazo)sulfides	C-N=N-S
Chlorites	O-Cl-O
Diazo compounds	CN$_2$
Diazonium salts	C-N$_2^+$
Difluoroamino compounds	N-F$_2$
Guanidinium oxosalts	N-C(=N$^+$H$_2$)-N
Halates	O-X-O$_2$
Haloacetylenes	C⦂C
N-Halogen compounds	N-X
Halogen oxides	O-X-O
Hydroxylaminium salts	N$^+$-OH
Hypohalites	O-X
Isoxazoles	C=N-O-C
Metal acetylides	C⦂C
N-Metal derivatives	N-Metal
Metal fulminates	C⦂N→O
Nitro compounds	C-NO$_2$
N-Nitro compounds	N-NO$_2$
Nitroso compounds	C-NO
N-Nitroso compounds	N-NO
Oximes	C=N-O
Ozone	O$_3$
Perchloryl compounds	C-Cl-O$_3$
Perchlorylamide salts	N-Cl-O$_3$
Perhalates	O-X-O$_3$
Peroxides	O-O
gem-Polynitroalkyl compounds	C-(NO$_2$)$_n$
Tetrazoles	N=N-NH-N
Triazines	C-N=N-N-C
Xanthates	C-N=N-S
Xenon-oxygen compounds	Xe-O$_n$

Specific examples of incidents reported with these chemicals are available.[4-6]

4 How Do You Assess the Risks?

4.1 Preliminary Investigations

It would be a great benefit if there were a fail-safe way of assessing the possibility of runaway reactions taking place, either by a formula that could be applied or by the use of a computer program. Computers programs do exist which claim to predict the reactivity of individual chemicals and mixtures of chemicals.[7-13] These programs may give some indication of the possibilities of reactivity but no more.

There is, however, a course of action which, if followed, should ensure that a reaction is carried out safely.

a) First, find out what is known about the reaction or related reactions from colleagues or from literature surveys.

b) If insufficient information is available, cautiously carry out a small scale preliminary reaction to assess the exothermic character and physical properties of the system and the products.

c) From the information gained in the small-scale experiment, the reaction can be scaled up in small stages. In this scale-up procedure there are certain points to remember:

   i) Control the temperature. Adequate provision for both heating and cooling the reaction must be made. If a certain amount of heat is required to initiate the reaction it may then need rapid cooling to maintain a safe reaction rate;

   ii) Give adequate thought to the proportions and concentrations of the reactants. The amounts of any potentially hazardous intermediate or by-product should be minimized by considering its mechanism of formation and subsequent reaction;

   iii) Make sure the reactants are pure. Impurities may catalyse the reaction;

   iv) Assess what order and at what rate the reactants should be added. The reaction may have an induction period before starting. Allowing a high concentration of reactants to be present in the reaction system after the induction period may lead to an uncontrollable reaction;

   v) Decide whether the reaction requires stirring and at what rate. Ineffective stirring can also lead to a build-up of concentrations of reactants.

4.2 HAZOP and HAZAN

There are two other assessment methods being increasingly used in laboratories to assess the risk of runaway reactions. These are the Hazard and Operability (HAZOP) study and Hazard Analysis (HAZAN).

HAZOP is used to identify hazards before an accident (or a runaway) occurs, while HAZAN is a technique for estimating the probability and consequences of a hazard and comparing them with a target or a criterion.

4.2.1   HAZOP — A What If? Study.

HAZOP is a method developed by ICI, and used widely in the UK chemical industry to find unanticipated dangers and help to control them.[14]

In a HAZOP study all the stages of a process or reaction are considered independently and in detail. All the possible hazards are identified in a systematic way, and steps are put in place to minimize their effect, bearing in mind the cost of any safety precautions required compared with the size of the hazard and the risk of that hazard occurring.

In larger organizations or when assessing larger, pilot scale reactions, the HAZOP study should be undertaken by a team of people who have the diverse knowledge required to cover all elements of the reaction systems and equipment. This allows ideas to be exchanged and debated to ensure, as far as possible, that all hazards are identified. It can also foster an atmosphere of hazard awareness throughout an organization as more workers become part of a HAZOP team.

In the laboratory, the team may be made up of a minimum of three people, *e.g.* the chemist whose reaction or process is being investigated, the safety officer, and someone from another laboratory. They all fulfil different but highly important roles, essentially by looking at the situation from different perspectives and different levels of knowledge.[15]

Initial studies should be carried out on existing experiments and apparatus to gain experience in the techniques before tackling more unknown situations.

In the laboratory, examples of the type of questions asked in a HAZOP study include:

Question — What happens if no water flows to a reflux condenser?

Consequences — It would lead to release of vapour, and the flask boiling dry.

Question — What if more catalyst or reagent is added to the reaction system?

Consequences — The reaction could overheat and lead to a runaway reaction.

The most common results from a HAZOP study are changes in equipment, materials or working methods, including:

- substitution with less toxic, flammable or reactive chemicals,

- moving to another laboratory with more appropriate facilities,

and not, as some fear, to increased costs of safety equipment.

HAZOP has also been used in laboratory design.[16]

4.2.2 HAZAN

HAZAN is a statistical method of estimating the hazard of an operation, and consists of three steps:

i) Estimating how often a particular incident or event will happen;

ii) Estimating the consequences to employees, the public and the environment, to the company or institution and any financial aspects;

iii) Comparing the results from i) and ii) with set targets in order to decide whether or not action is necessary to minimize the consequences or whether the hazard can be ignored.

HAZAN is of less use to the small scale laboratory experiment and more useful in pilot plant scaled-up reactions. It is also a technique that requires a certain amount of guidance from someone with experience of the technique.

Both HAZOP and HAZAN methods are discussed at length with respect to the process industry in 'HAZOP and HAZAN: Identifying and Assessing Process Industry Hazards'.[17]

5 Precautions for Performing Laboratory Reactions

After the preliminary investigations, literature searches, and pilot experiments, the final assessment that must be carried out is 'can the reaction be carried out safely?' The answer to this question depends to a great extent on whether the reaction has been assessed as hazardous and what safety equipment is available to the researcher. The following discussion is an attempt to simplify this process.

If after thorough risk assessment it has been concluded that the reaction is not a runaway hazard and the reactants and products are non-toxic, then a small-scale reaction can be undertaken on a laboratory bench, facing the laboratory wall, behind a safety screen.

It should also be noted that the minimum requirements for safety equipment in a laboratory are a laboratory coat and safely spectacles or face visor.

Large-scale reactions with non-toxic reactants or products, or small-scale reactions with toxic reactants or products, should be carried out in a fume cupboard with the sash lowered. If the chemicals involved also pose a threat of causing a runaway reaction, then it is prudent to have a fire extinguisher readily at hand.

For reactions involving highly toxic compounds in any quantity, the protection afforded by a fume cupboard should be supplemented with the availability of a respirator or breathing apparatus.

Finally, if there is the possibility of the reaction or products undergoing explosive decomposition, or the reaction requires high pressures, then that reaction should be carried out in an isolated cell.

Table 1 is based on information contained in The Royal Society of Chemistry (RSC) publication 'Hazards in the Chemical Laboratory'[5] and summarizes the precautions quoted above. Further information on isolated cells is available,[18] as is guidance on laboratory fume cupboards.[19]

Table 1 *Minimum precautions for performing laboratory reactions*

Reaction Scale	Reactants/Products	Precautions
Small	Not toxic	Laboratory bench behind safety screen
Large/ Small	Not toxic/ Toxic	Fume cupboard sash lowered
Any	Highly toxic, especially cylinder gases	Fume cupboard sash lowered - respirator available
Any	Explosive decomposition possible or high pressure reaction	Isolated cell

Another method of evaluation of risk assessment and control determination is published in the RSC publication 'COSHH in Laboratories'.[20] This technique takes into account the intrinsic hazard of the substances under assessment and the potential for exposure based on the quantity of the substance being used, its physical characteristics, and the characteristics of the reaction methodology. Combination of the hazard and exposure potential leads to the calculation of a risk factor and hence to containment regimes required for the reaction. These containment regimes are:

i)   Open bench;

ii) Fume cupboard (or other specially vented area);

iii) Special facilities.

It should be stressed that if the appropriate containment and protective measures are not available at the company or institution, then the reaction should not be attempted until those protective measures and containment are installed.

6 What Information Sources Are Available to Identify Runaway and Hazardous Reactions?

Two sources of information are available on the hazardous and runaway potential of a chemical reaction: standard reference books[4,6] and current awareness services.

Books

Many books contain information drawn from various sources on the reactivity of chemicals, singly or in combination with others. These can be part of chemical safety data sheets as required under EC legislation and contain a small section on the reactive chemical hazards of the chemical, or there are a few compilations which deal solely with reactive chemical hazards.

Current awareness services

These services scan the literature for various types of information and publish printed or computer-readable, *e.g.* online, files periodically.

In trying to find out the potential hazards associated with a particular reaction, a combination of both these information sources is required to cover adequately the possibility of someone else having identified a hazard of which you, as the person carrying out the experiment, should be aware.

Books are a snapshot of the situation, the knowledge at the particular time when they are published. However, in any book-publishing cycle there tends to be a delay between updating information and bringing out new editions. This is where the current awareness services are so useful. They effectively fill in the gap between the time a book was published and the present time. Therefore by using a combination of both books and current awareness sources all the required information should be identified.

In this scheme of books and current awareness services there are some important sources that should be consulted.

6.1 Standard Reference Books on Hazardous and Runaway Reactions

6.1.1 *Handbook of Reactive Chemical Hazards, 1990*[4]

This book is a collection of the reactive chemical hazards reported for over 4600 chemicals and classes of chemicals. It aims to give access to a wide and up-to-date selection of documented information to research students, practising chemists, safety officers and others concerned with the safe handling and use of reactive chemicals. It allows ready assessment of the likely potential for reaction hazards which may be associated with an existing or proposed chemical compound or reaction system.

The 4th edition of 'Bretherick', published in 1990, includes all information available to the author by April 1989 on the reactive hazards of individual elements or compounds either alone or in combination. This information comes from a wide number and variety of sources, and includes primary journals that specialise in the publication of safety matters and textbooks specialising in synthetic methods. Major encyclopedic collections consulted are:

Comprehensive Treatise on Inorganic and Theoretical Chemistry,
J. W. Mellor, Longmans Green, London, Vol. 1 repr. 1941, Vol. 16. publ. 1937; and isolated supplementary volumes up to Vol. 8 Suppl. 3, 1971,

Houben-Weyl — Methoden der Organischen Chemie,
ed. E. Müller, Georg Thieme, Stuttgart, 4th Edition, 16 Vols in parts, 1953 to date.

Secondary sources or abstracting publications are also used, *e.g.*

Section 50 of Chemical Abstracts — Chemical Hazards, Health & Safety,

Universities' Safety Association Safety News,

Loss Prevention Bulletin (IChemE),

Laboratory Hazards Bulletin (RSC),

Chemical Hazards in Industry (RSC).

Included in the chemicals listed are 560 gases and liquids with flash points below 25 °C and/or autoignition temperatures below 225 °C.

The individual items in the book contain the chemical name, molecular formula and structure. Information then follows on the hazards identified for that chemical or the hazard or incident and a reference for that information. Some information derived from private communications is included.

The information is indexed by chemical name, synonym and CAS Registry Number. There is also a section covering information on classes of chemicals and topics, structured in a similar way to that on individual chemicals. This section is indexed by class, group and topic.

Example 1 shows an item from 'Bretherick'.

Example 1

4322. Potassium permanganate
   [7722-64-7]   KMnO₄
                              KMnO₄

Acetone, *tert*-Butylamine   MRH   Acetone 2.76/8

1. Kornblum, N. *et al., Org. Synth.*, 1973, Coll. Vol. 5, 847
2. Kornblum, N. *et al., Org. Synth.*, 1963, **43**, 89
3. Haynes, R. K. *et al., Lab. Equip. Digest*, 1974, **12**(6), 98; private comm., 1974

In the reprinted description [1] of a general method of oxidising *tert*-alkylamines to the corresponding nitroalkanes, a superscript reference indicating that *tert*-butylamine should be oxidised in water alone, rather than in acetone containing 20% of water, is omitted, although it was present in the original description [2]. This appears to be important, because running the reaction in 20% aqueous acetone led to a violent reaction with eruption of the flask contents. This was attributed to caking of the solid permanganate owing to inadequate agitation, and onset of an exothermic reaction between oxidant and solvent [3].

It is worth noting that the information in this publication is now available in a PC form[21] which makes it slightly easier to use than the book. In the book the information is only held once, *i.e.* in the previous example there would only appear a cross reference at Acetone and *tert*-Butylamine pointing the reader to Potassium permanganate. The PC product is similarly structured, although it is easy to highlight the relevant cross reference on the computer screen which then instructs the software to display the primary item where the information is held. This is much more user-friendly than the printed publication.

6.1.2 The Fire Protection Guide on Hazardous Materials, 1991[6]

This compendium contains three main sections.

i) Fire Hazard Properties of Flammable Liquids, Gases, and Volatile Solids

This section alphabetically lists the fire hazard properties of over 1300 flammable substances by name. The values selected are representative figures suitable for general use. Data reported include flash point, ignition temperature, flammable limits, specific gravity, vapour density, boiling point and extinguishing methods.

ii) Hazardous Chemicals Data

This section contains fire, explosion and toxicity hazards data on over 400 chemicals. The chemicals appear alphabetically and the list contains synonyms cross-referred to the preferred name for that chemical. Listed are a description of the chemical, its fire and explosion hazards, life hazards, personal protection, fire fighting phases, usual shipping containers and advice on storage.

iii) Manual of Hazardous Chemical Reactions

This section includes information on over 3500 mixtures of two or more chemicals reported to be potentially dangerous in that they may cause fires, explosions, or detonations at ordinary or moderately elevated temperatures. It is arranged alphabetically, and chemicals and synonyms are cross-referred to the preferred name.

The stated purpose of the manual is to 'bring to chemists and other users of chemicals a compilation of recorded experience with chemical reactions that have potential for danger'. Each item contains a chemical name or class followed by, where appropriate, a molecular structure. Below this is the chemical or class of chemical which, with the first chemical, has been reported to lead to a hazardous reaction. A description of that reaction is then given and a reference to the original source of data is quoted.

Example 2 shows an item from this section.

Example 2

BUTADIYNE HC⦂CC⦂CH

Air	In the preparation of diacetylene by adding 1,4-dichloro-2-butyne to 10% sodium hydroxide and a little dioxane at 100°C, no difficulty was experienced if the free diacetylene was collected at minus 25 °C and held below that temperature. At temperatures above minus 25 °C explosions would occur. *Rutledge*, pp. 134-135 (1968).

This section, whilst being a very useful source, suffers the same draw back in cross referencing as 'Handbook of Reactive Chemical Hazards'.

6.2 Laboratory Hazards Bulletin — A Current Awareness Service

The Royal Society of Chemistry publishes a monthly current awareness bulletin called *Laboratory Hazards Bulletin*.[22] This bulletin contains references to the literature on hazards and health and safety for workers in laboratories of all types. It covers a wide range of information including reactive hazards, biological effects of chemicals, animals and microorganisms, legislation and standards, articles on laboratory design, management and practice, waste disposal, fume cupboards, and occupational health and hygiene. Appendix 1 lists the individual sections in the Bulletin and their scope and content.

The information covering reactive hazards is selected from over 150 journals which are regularly scanned for information on particular subjects relevant to the laboratory environment. Appendix 2 lists the main journals from which items have appeared over the past few years.

Items in the *Laboratory Hazards Bulletin* consist of 4 parts: i) Title, ii) Bibliography, iii) Abstract, iv) Indexes.

The *title* is that of the original article or its English translation. In the case of a translated title, the original language title is also included. Where the title of the original article is not the interesting aspect on which the abstract is based, the title of the article is changed to a relevant one for the subject covered in the abstract. The original title of the article is also given to enable the original article to be easily located in the literature. These articles may describe the preparation of a new product where a reactive chemical hazard is highlighted within the experimental section and not elsewhere in the article.

The *bibliography* indicates the source of information. For journal articles, the authors and their affiliation (company or institution) and contact address, the journal where the article appeared, its year of publication, volume, issue, and page ranges are all quoted. For a book, the author, date of publication, publisher and ISBN are given. For other documents a contact name and address is given.

The *abstract* provides a clear and concise summary of the main findings of the original article which are relevant to the hazards of chemicals. Data are reported from the original article when relevant.

There are two *indexes* in the printed publication — a subject index and a chemical index. The subject index consists of controlled vocabulary terms so that all similar articles are indexed under the same term, *e.g.* items about fume cupboards or hoods appear under the heading of FUME CUPBOARD and <u>not</u> FUME HOOD or any other term. Similarly the chemicals index is generated in such a way that all identical chemicals are indexed under the same chemical name, *e.g.* chloroform appears at CHLOROFORM and <u>not</u> TRICHLOROMETHANE.

Information in *Laboratory Hazards Bulletin* is also available as an online file *via* four online vendors under the name Chemical Safety NewsBase.[22] It is, therefore, possible to conduct online searches of the information very easily if you have a PC connected to a modem.

The database has all the features of a normal online database where it is possible to search for a particular word within an item, but the main ways of searching are *via* chemical name and CAS registry number (the third index in the online file) for identification of a particular chemical, or by using the extensive subject index of the database.

Examples 3 and 4 are two items reproduced with permission from the RSC *Laboratory Hazards Bulletin*.

Example 3 — Item from Laboratory Hazards Bulletin

EXPLOSION WITH LITHIUM PERCHLORATE IN DIELS-ALDER REACTION. *Silva, R. A. (Calif. State Univ., Northridge, CA, USA). Chem. Eng. News, 21 Dec 1992, 70(51), 2.*

A violent explosion occurred during the use of lithium perchlorate in a Diels-Alder cycloaddition reaction. 1 g of cyclooctatetraene (COT), 1 g of dimethylacetylene dicarboxylate and 5 g of lithium perchlorate were dissolved in 9.5 ml dry ether. The mixture was brought to flux in a thermowell heater and left in a fume hood, where it exploded after 90 minutes. COT and lithium perchlorate can react violently under certain conditions, so extreme caution is recommended when they are combined.

Example 4 — Item from Laboratory Hazards Bulletin

THERMAL HAZARDS WITH *TERT*-BUTYL NITRITE
Lopez, F. (Schering AG, Berlin, Germany). Chem. Eng. News, 21 Dec 1992, 70(51), 2.

During a hazard assessment it was found that *tert*-butyl nitrite can decompose on impact and thermally. Smoke and soot formed after a standard falling hammer test. DSC scanning showed that this compound undergoes strongly exothermic decomposition starting at around 110 °C. Similar hazards have been reported with analogous compounds. *Tert*-butyl nitrite is commonly used in organic synthesis. It can be replaced with a saturated aqueous solution of potassium nitrite. If usage is continued it is advisable to consult a safety expert to ensure appropriate safety measures are used.

Example 3 reports a violent explosion in a particular reaction highlighting the extreme caution required in carrying out the procedure.

Example 4 reports the thermal and shock sensitivity of *tert*-butyl nitrite and suggests a safer alternative.

Both of these incidents reported from 1992 would not have appeared in the standard reference books.[4,6] A combination of reference books and updated abstract literature provides the best means of keeping an experimental scientist informed about hazards.

6.3 Other Sources of Abstract Information

Other sources include:

* CA Selects — which are subsets of Chemical Abstracts in various areas including:

   Carcinogens and mutagens
   Chemical hazards, health and safety
   Flammability
   Occupational exposure and hazards,

* Laboratory Safety Supply Abstracts, which is a customer service available from Laboratory Safety Supply International Division, PO Box 1441, Slough SL1 8JS, UK.

By using these sources of information in combination with other more general sources such as Chemical Abstracts, and safety data sheets, enough information about an 'unknown' reaction can usually be found for risk assessments, and thus avoid runaway reactions and, more importantly, injury to the researcher.

7 Conclusion

In my experience all chemists regard themselves as being 'good' chemists. There seems to have been a tendency in the last few years for chemists to think that 'good' chemists do not have runaway reactions. However, there is no evidence to support this supposition. What it does mean is that chemists do not seem so willing to reveal when they are involved in a hazardous incident or have a runaway reaction.[23] It is quite noticeable that the number of items on reactive chemical hazards within the literature has declined over the last few years. Items appearing in *Laboratory Hazards Bulletin* - Section 1 (Fires and Explosions) numbered 228 in 1984 but only 140 in 1992.

It should be in every chemist's mind that if, and when, they experience a hazardous incident, then perhaps someone else has had a similar experience and has not reported it. How would you regard that person, especially if the incident had caused you injury? The whole chemical community profits from the effective reporting of chemical hazards, and there is **no** excuse not to report these incidents.

8 References

1. T. S. Carswell, *J. Am. Chem. Soc.*, 1931, 2417 (quoted in L. Bretherick, 'Handbook of Reactive Chemical Hazards, 4th Edition', Butterworths, London, 1990, 710).

2. P. W. Atkins, 'Physical Chemistry', 4th Edition, Oxford University Press, Oxford, 1990, Chapter 26, pp. 792-794.

3. MCA Case History No. 1482 (quoted in L. Bretherick, 'Handbook of Reactive Chemical Hazards, 4th Edition', Butterworths, London, 1990, pp. 607-608).

4. L. Bretherick, 'Handbook of Reactive Chemical Hazards, 4th Edition', Butterworths, London, 1990.

5. 'Hazards in the Chemical Laboratory', Fifth Edition, ed. S.G. Luxon, Royal Society of Chemistry, Cambridge, 1992.

6. 'The Fire Protection Guide on Hazardous Materials, 10th Edition', National Fire Protection Association, MA, USA, 1991.

7. T. Yoshida, *Kogyo Kayaku*, 1977, **38**, 85 (*Chem. Abstr.* **88**, 95673).

8. T. Yoshida, (*Sogo Shikensho Nenpo Tokyo Daigaku Kogakubu*), 1978, **37**, 235 (*Chem. Abstr.* **92**, 168382).

9. T. Yoshida, *Kenkyu Hokoku - Asahi Garasu Kogyo Gijutsu Shoreikai*, 1979, **35**, 187 (*Chem. Abstr.* **93**, 188632).

10. T. Yoshida, *Anzen Kogaku*, 1983, **22**, 12 (*Chem. Abstr.* **99**, 7923).

11. T. Yoshida, *Kogyo Kayaku*, 1984, **45**, 66 (*Chem. Abstr.* **102**, 169208).

12. T. Yoshida, 'Safety of Reactive Chemicals', Elsevier, Amsterdam, 1986.

13. 'CHETAH, ASTM Chemical Thermodynamics and Energy Release Evaluation Program', American Society for Testing Materials, Philadelphia, release 4.4, 1989.

14. T. A. Kletz, *Inst. Chem. Eng. (UK)*, 1984.

15. M. J. Pitt, *J. Chem. Educ.*, 1987, A44-45.

16. R. E. Knowlton, *R&D Management*, 1976, 1.

17. T. Kletz, 'HAZOP and HAZAN: Identifying and Assessing Process Industry Hazards, 3rd Edition', IChemE, 1992.

18. W. G. High, 'The Design of a Cubicle for Oxidation or High-Pressure Equipment', *Chem. Ind.*, 1967, 899; A. L. Glazebrook, 'Safety in the Study of Chemical Reactions at High-Pressure', Tech. Bull. No. 100, Autoclave Engineers, Erie, 1974.

19. 'Guidance on Laboratory Fume Cupboards', Royal Society of Chemistry, London, 1990.

20. 'COSHH in Laboratories', Royal Society of Chemistry, London, 1989.

21. 'Reactive Chemical Hazards Database', Butterworth-Heinemann, London, 1991.

22. 'Laboratory Hazards Bulletin', Royal Society of Chemistry, 1981-1993. Also available online as: Chemical Safety NewsBase *via* DIALOG, Data-Star, STN and Orbit.

23. P. Urben, *Chem. Ind.*, 1991, 143.

APPENDIX 1

LABORATORY HAZARDS BULLETIN

SECTION SCOPE NOTES

The following details describe the sections within *Laboratory Hazards Bulletin* and the scope of each of these sections.

1. Fires & Explosions
News items on actual incidents, plus articles about prevention, protection and risk assessment. Runaway reactions.

2. Hazardous waste management
General aspects of waste management and disposal including specific techniques; incineration, biological or chemical treatment, landfill and recycling, *etc.* Emission controls.

3. Storage & Transportation
General aspects of transportation and specific types: road, rail, maritime, air and in-plant. General aspects of storage, and specific types, *e.g.* underground storage tanks.

4. Leaks, Spills & Unplanned Releases
News items on leaks and spills plus articles on prevention techniques and safe practices.

5. Animal & Microbiological Hazards
Studies and news items on biotechnology, AIDS and HIV virus, hepatitis and other viruses, bacteria, general infection control (*e.g.* sharp injury, tissue/specimen handling).

6. Carcinogens & Mutagens
Studies in humans, laboratory animals and cell cultures resulting in cancer and genetic changes.

7. Reproductive hazards
Studies of reproductive disorders in men, women and their offspring plus relevant studies in laboratory animals and cell cultures.

8. Allergy & Irritants
Studies in humans of internal and external dermal disorders of allergic and irritant origin (dermatitis, urticaria, asthma, rhinitis, *etc.*), plus relevant studies in laboratory animals and cell cultures.

9. General & Miscellaneous Biological Effects
Studies of chemicals not yet recognised to be toxic and putative toxic chemicals in humans, laboratory animals and cell cultures. Aspects of biotechnology, and toxicity testing and hazard evaluation of chemicals.

10. General Industrial Hazards
Radiation (ionizing, laser, microwave, X-ray, etc), electrical hazards, noise, slipping, tripping & falling, lifting, heating & ventilation and housekeeping.

11. Legislation
Legislation and standards from throughout the world, with emphasis on the EC, USA and Australia, plus news, discussions and litigations.

12. Precautions
Accident prevention, safety management, plant process & laboratory design, non-emergency equipment, classification, packaging & labelling, operating procedures/systems, process monitoring & leak detection, protective clothing & equipment, engineering controls, maintenance & repairs, alarm & emergency relief systems, emergency planning, accident investigation & statistics, computer applications, education & training.

13. Occupational Health, Hygiene & Monitoring
General occupational health, hygiene and nursing, occupational disease diagnosis, occupational health and chemical exposure records, mortality, morbidity, accident and injury statistics, first aid, and monitoring and biological monitoring.

14. Forthcoming Events
Courses, conferences, symposia, exhibitions.

15. Publications, Organisations, etc.
All films, videos, posters, data sheets, books, conference proceedings, technical reports, reviews, news of new journals, databases, databanks, organizations, societies, advice and consultancy services.

Articles of interest to more than one are cross-referred between sections.

APPENDIX 2

JOURNALS CITED IN LABORATORY HAZARDS BULLETIN

AAOHN J.
Adv. Mod. Environ. Toxicol.
AIChE Symp. Ser.
AIDS
AIDS Newsl.
Am. Fam. Phys.
Am. Ind. Hyg. Assoc. J.
Am. J. Emerg. Med.
Am. J. Epidemiol.
Am. J. Forensic. Med. Pathol.
Am. J. Gastroenterol.
Am. J. Hosp. Pharm.
Am. J. Ind. Med.
Am. J. Obstet. Gynecol.
Am. Lab. (Fairfield, Conn.)
Anaesthetist
Anal. Chem.
Anal. Proc. (London)
Anal. Sci.
Angew. Chem., Int. Ed.
Ann. Chir. Main Memb. Super.
Ann. Dermatol. Venereol.
Ann. Emerg. Med.
Ann. Intern. Med.
Ann. N.Y. Acad. Sci.
Ann. Occup. Hyg.
Appl. Environ. Microbiol.
Appl. Microbiol. Biotechnol.
Appl. Occup. Environ. Hyg.
Appl. Optics
Appl. Organomet. Chem.
Arbeidervern
Arbeitsmed., Sozialmed.,
 Praeventivmed.
Arch. Environ. Health
Arch. Mal. Prof. Med. Trav. Secur.
 Soc. Arch. Pathol. Lab. Med.
Arch. Toxicol.
Arthritis Rheum.
ASTM Spec. Tech. Publ.

ASTM Spec. Tech. Publ. Bioproces
 Saf.
At the Centre
Aust. J. Chem.
Aust. Stand.
Biologicals
Biotec (Brescia)
Biotechnol. Bioeng.
Br. J. Ind. Med.
Br. Med. J.
Brain
Brain Res.
BSI News
Bull. Environ. Contam. Toxicol.
Burns
Cah. Notes Doc.
Can. Med. Assoc. J.
Canada Communicable Dis. Rep.
Cancer
Cancer Lett. (Shannon, Irel.)
Centrifuge User
Cesk. Hyg.
Chem. Br.
Chem. Eng. News
Chem. Ind. (London)
Chem. Prog. Bull.
Chem. Week
Chemosphere
Chest
Chim. Oggi
Clin. Lab. Manage. Rev.
Clin. Radiol.
Clin. Sci.
Clin. Toxicol.
Collect. Czech. Chem. Commun.
Contact Dermatitis
CRC Crit. Rev. Toxicol.
Crit. Care Med.
Daily Hazard
Dangerous Prop. Ind. Mater. Rep.

Derm. Beruf. Umwelt
Derm. Berufumwelt
Drug Dev. Res.
Drug. Saf.
EHP, Environ. Health Perspect.
ENDS Rep.
Environ. Health
Environ. Manage.
Environ. Mol. Mutagen.
Environ. Profess.
Environ. Prot. Bull.
Environ. Res.
Environ. Sci. Technol.
Environ. Toxicol. Chem.
Epidemiology
Eur. Chem. News
Eur. Clin. Lab.
Eur. Microbiol.
Eur. Resp. J.
Eur. Safety Newsl.
Euronews
Farb. Lack
Fed. Regist.
Filtr. Sep.
Fire Prev.
Fire Surv.
Food Chem. Toxicol.
Fresenius. J. Anal. Chem.
Fundam. Appl. Toxicol.
Gig. Sanit.
Hazards
Health Educ. News
Health Phys.
Health Saf. Environ. Bull.
Health Saf. Inf. Bull.
Health Saf. Work
Hereditas
Histochem. J.
HSC Newsl.
HSE Transl.
Hum. Exp. Toxicol.
Hung. J. Ind. Chem.
IC, Infect. Control
IES News
IIRSM Newsl.

Ind. Chem. Libr.
Ind. Environ.
Ind. Health
Indoor Environ.
Infect. Control Hosp. Epidemiol.
Infection
Ing. Quim. (Madrid)
Inhalation Toxicol.
Inorg. Chem.
Inst. Chem. Eng. Symp. Ser.
Int. Arch. Occup. Environ. Health
Int. Biotechnol. Lab.
Int. Biotechnol. Lab. News
Int. Environ. Saf. News
Int. J. Cancer
Int. Lab.
IRPTC Bull.
J. Air Waste Manage. Assoc.
J. Allergy Clin. Immunol.
J. Am. Acad. Dermatol.
J. Am. Coll. Toxicol.
J. Am. Med. Assoc.
J. Appl. Bacteriol.
J. Burn Care Rehabil.
J. Cancer Res. Clin. Oncol.
J. Chem. Educ.
J. Chem. Soc., Chem. Commun.
J. Chromatogr. Biomed. Appl.
J. Emerg. Med.
J. Environ. Sci. Health, Part A
J. Fluorine Chem.
J. Forensic Sci.
J. Fr. Ophtalmol.
J. Hazard. Mater.
J. Hosp. Infect.
J. Immunol. Methods
J. Invest. Dermatol.
J. Loss Prev. Process Ind.
J. Microsc.
J. Occup. Health Saf. Aust. New
Zealand
J. Prim. Prev.
J. Radioanal. Nucl. Chem.
J. Radiol. Prot.
J. Soc. Occup. Med.

J. Toxicol. Environ. Health
J. Toxicol., Clin. Toxicol.
J. UOEH
JAMA, J. Am. Med. Assoc.
Janus
JAT, J. Appl. Toxicol.
JOM, J. Occup. Med.
Jpn. J. Ind. Health
Jpn. J. Pharmacol.
Kagaku to Kyoiku
Lab. Manage. Rev.
Lab. News
Lab. Pract.
Lab. Prax.
Labor-med.
Labour Research
Lancet
Louvain Med.
Med. Clin. North Am.
Med. Hypotheses
Med. Lab. Sci.
Med. Lav.
Metallkurier
Morbid. Mortal. Wk. Rep.
Mutat. Res.
Nature (London)
Ned. Chem. Ind.
Neurotoxicol. Teratol.
New Engl. J. Med.
New Sci.
Newsl. Natl. Board Occup. Saf. Health
 (Sweden)
Newsl., Swed. Work Environ. Fund
Noticias de Seguridad
Occup. Hazards
Occup. Health
Occup. Health Bull.
Occup. Health Ontario
Occup. Health Rev.
Occup. Health Saf.
Occup. Med.
Occup. Med.: State of the Art Rev.
Occup. Saf. Health
Occup. Saf. Health Reporter
Off. J. Eur. Commun.

Ophthalmology
Organometallics
Otolaryngol. Clin. North Am.
Pharm. Technol
Pharmacol. Biochem. Behav.
Pharmacol. Toxicol. (Copenhagen)
Plant/Oper. Prog.
Plast. Reconstr. Surg.
Pol. J. Occup. Med.
Pol. J. Occup. Med. Environ. Health
Poll. Eng.
Polym., Paint Colour J.
Postgrad. Med.
Prac. Lek.
Prev. au Travail
Process Ind. J.
Processing
Prof. Saf.
Promosafe
Psychiatr. Prax.
Public Health Rep.
Quim. Nova
Radiat. Prot. Dosim.
Recomb. DNA Tech. Bull.
Reprod. Toxicol.
Res. Dev.
Res. Technol. Internatl.
Risk Anal.
RoSPA Bull.
S. Afr. Med. J.
Saf. Dig. Univ. Saf. Assoc.
Saf. Health (Chicago)
Saf. Health Japan
Saf. Health Pract.
Saf. Man. Newsl.
Saf. Manage.
Saf. Manage. (London)
Saf. Manage. (Veiligheidsbestuur)
Saf. Managers Newsl.
Saf. Matters
Saf. Pract.
Safe. Manage.
Safety Managers Newsl.
Scand. J. Work, Environ. Health
Sch. Sci. Rev.

Semicond. Int.
Sichere Chemiearb.
South. Med. J.
Staub - Reinhalt. Luft
Strategies Molecular Biol.
Synthesis
Teratog., Carcinog., Mutagen.
Tetrahedron
The Chemical Engineer
The Guardian
The Scientist
Toxic Subst. Bull.

Toxicol. Appl. Pharmacol.
Toxicol. Ind. Health
Toxicol. Lett.
Toxicol. Pathol.
Toxicology
Waste Manage.
What's New Process
Wien. Med. Wochenschr.
Wiss. Umwelt
Workers Health Internatl. Newsl.
Z. Gesamte Hyg. Ihre Grenzgeb.
Zh. Org. Khim.

CHAPTER 12

Laboratory Design: Determining and Meeting Clients' Requirements

C. BLOCH

1 Introduction

Laboratory space is among the most expensive and hazardous of all built spaces, and is also one of the most difficult to design in order to fulfil the requirements of those who work within it. Risks associated with innovation lead many designers, whether they are architects or engineers, to reproduce without question previous designs, and replicate features within one building, such as a typical bench layout or a large modular laboratory, irrespective of the different scientific activities which may take place. About 300,000 people work, teach and learn in laboratories in the UK every day. Because of the hazards associated with a laboratory environment, it is essential that the best possible facilities are designed which also reconcile safety with productivity and cost with value.

Architects are limited in their ability to design innovatively for laboratory clients compared with other projects. While architects may have gained experience as users of hotels, offices, hospitals, shops or even prisons, they will almost certainly not have experienced laboratory work, and will therefore be inhibited and inevitably restricted by their lack of understanding in suggesting innovative approaches. Consequently in matters of layout and detail, architects must rely almost exclusively upon the personal preferences of the client or his or her representative.

Scientists by contrast are notorious for asking architects to provide more of the same or to repeat what has been built elsewhere. The attitudes and experiences of architects and scientists are rendered more complicated by each regarding their own profession as having a monopoly on innovation. Paradoxically it is the scientist with the truly innovative approach who is always more valuable to a company for his or her scientific work than to be spared for such mundane tasks as briefing architects. To compound further the difficulties of the two vocations establishing common ground, each has a tendency to hide behind its own distinctive and protective professional jargon.

This preamble sets the background and illustrates some problems in briefing.[1] The ultimate success of a laboratory design can be guaranteed more by a fluent and comprehensive briefing regime than perhaps any other parameter. If the right brief is received then design, procurement and construction are relatively straightforward technical tasks. But without an accurate brief, all these subsequent stages risk being progressively corrupted, resulting inevitably in degrees of failure in achieving the traditional objectives of cost, quality and time.

This chapter discusses aspects of briefing, brief gathering, and the problems of briefing as an effective technique for determining clients' needs and translating them into completed buildings.

2 Defining the Project

Objectives must not only be stated at the outset, but must be presented in a way that can be commonly understood by clients, users, and designers. This must include a verification for the project's need, justification for its form, *i.e.* whether it is a new building or a refurbished one, its life expectancy, and the requirements for flexibility and occupancy.

Project objectives should address cost, quality, time — and risk; the stated project objective should establish acceptable levels of risk in the design and particularly in the procurement of the project.

Major corporate clients, especially in the pharmaceutical sector, now define not only project objectives but also a mission statement. The latter deals particularly with unquantifiable aspects of employment such as corporate culture, quality of work and of working, and employment sociology or the way in which the organization of building spaces should reflect hierarchy. Inter-personal relationships may seem slightly frivolous, but if the architect is given an understanding of the desired relationship between different staff levels, it will be reflected to the benefit of all in the layout and planning of the building. One company or institution may have open plan offices, a single cafeteria, and glass-walled meeting rooms, while another may have a high proportion of cellular offices, executive dining rooms and panelled enclaves of meeting and board rooms. Grades of luxury and amenities included within a brief, and hence in the building, reveal the corporate image both to employees and visitors. Such notions are susceptible to change, however, and a level of flexibility must be provided in the building's structure, fabric and servicing for the inevitable shifts in emphasis. Just as open plan offices are a fading fashion, so the workplace sociologist's dream of the communal photocopier and coffee station, over which erudite minds achieve new heights of serendipity, synergy and symbiosis, could also vanish.

Briefing is therefore vital to the success of a project, and must be comprehensive and cover more than mere technical and spatial requirements. As a quantifiable measure of its importance, it is salutary to note that some 15% of an architect's fee, and hence his or her time and cost, is normally allocated to the development of a brief.[2]

It is important also to distinguish between a client's *request*, and a client's *requirements*. A client's request is in fact a brief which has already been prepared by the client before involving the architect, and is a technique used if, for example, a design competition is envisaged and all competitors are working to the same criteria. It should be the role and the duty of architects to challenge a brief in all its aspects.

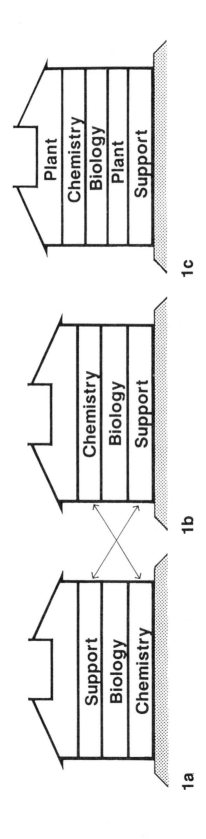

Figure 1 *The evolution of the assembly of a multi-disciplinary pharmaceutical laboratory*

For example, for a recently completed large pharmaceutical laboratory in the UK[3] the client required a multi-disciplinary building comprising, in approximately the same proportions, chemistry laboratories, biology laboratories, and support services.

The client requested that the three components of the building should be layered in the order shown in Figure 1a, with the chemistry laboratories on the ground floor (for security and communications reasons), and with the support services on the top floor, also for security. The building design evolved through challenge by the architects of the client's requests, into one which met their demands in a quite different way, which did not diminish the operational efficiency of the building, and which offered considerable economies in construction. The design process had in effect added value without detracting from the client's requirements.

The resultant layering, shown in Figure 1b, is an inversion of the original request, and was influenced by two considerations:

i)   Discharging the 200 fume cupboards needed for the chemistry floor would have consumed otherwise usable space;

ii) The security requirements for the support services floor could be satisfied equally as well on the ground floor.

Figure 1c shows the insertion of mechanical and electrical plant into the final arrangement for the building (Figure 1b), and illustrates that plant volumes in a highly-serviced building can account for more than half the building volume (and often more than half the building's cost).

3 The Clients' Representative

One of the difficulties facing architects is the actual identification of their client. Setting aside the name on the commissioning agreement, reality and experience show that the client — those who have to approve and/or sponsor the project — can be a combination of:

* Research representatives;

* Actual and potential users of the building;

* Financial representatives;

* Statutory controllers (*e.g.* planning and building regulators, fire officers);

* The wider public upon whose sensibilities the building will be imposed; and

* The architect's own professional peers and superiors.

Architects usually require from the client organization a single representative who will brief them and act as the single source for instructions and decisions. Selection of the procurer is perhaps the most important appointment which the organization can make, as it requires considerable levels of knowledge, experience, and qualities described thus:

'Normally and preferably appointed from within the company or institution, the procurer must have, in addition to an open and challenging mind and natural leadership qualities: (i) knowledge of research and of research work and a broad but not necessarily detailed knowledge of technical building requirements such as piped and wired services, air design and management, and principles of planning for safety; (ii) a clear understanding of corporate or work culture and objectives, particularly the relative priorities attached to the traditional but frequently mutually exclusive criteria of cost, quality and time; and (iii) the authority to make or obtain quick decisions, as delays to design work through indecision will invariably cost time and money.'[4]

The importance of a single point of representation and authority on the client's side cannot be sufficiently stressed. There is a developing trend in the procurement of laboratory, and indeed other major building projects, to appoint initially a third party project manager, usually a Chartered Surveyor but often a Chartered Architect, who assumes responsibility not only for determining and writing the brief, but also for appointing the design and construction teams, usually for a fee of between 1% and 2% of the project's capital cost. There are inherent risks in this approach, as the organization relies almost entirely on a third party, perhaps with little experience of the methods and culture of its work, to reconcile the often contradictory pressures on a project.

4 Establishing Common Ground

4.1 Establishing a Dialogue

In all projects, a compromise must be struck between the 're-invention of the wheel' or the unchallenged repetition of a previous design and the use of a tried, tired and tested solution for which it is thought there is no realistic alternative. Indeed we may challenge the whole notion of the laboratory as a separate type of building, and consider whether it is not in fact a factory environment with offices attached, or *vice versa*.

The laboratory has evolved during the twentieth century as the nature of scientific investigation has matured: 'The growth and changes which science has experienced in this century alone contradict the stereotyped image of the lone scientist in a white coat pouring vaporous chemicals from one test tube into another in a garret laboratory.'[5]

Unwitting intransigence referred to earlier is a view shared by others: 'Even for experienced (laboratory designers) there is a natural tendency to repeat what works rather than to test new ideas.... It is difficult for laboratory users to imagine facilities they have not experienced. Furthermore, scientific people, by their very nature and training, may unwittingly obscure attempts to understand laboratories and how they work by cloaking information in scientific jargon and tradition.'[6]

Errors in design can occur as a consequence of errors in communication. A user thinks in terms of functional requirements based on his or her own experience, and will translate these into spatial and physical needs. A designer will think in terms of physical requirements. In the design of a radiation laboratory, for example, the designer will consider its location relative to other parts of the building, the access and security arrangements, storage requirements, and decontamination and shower areas, and will translate these facets into physical dimensions and specifications for construction. The user on the other hand will consider the actual work programme to be carried out in the laboratory, and assume this will be similar to a current workload. It is seldom clear who has responsibility for asking whether that particular programme has to be contained within a dedicated radiation laboratory, what the laboratory's use is likely to be in five years time and if the work programme could be contracted out instead of constructing a dedicated building.

Such queries on the classification of laboratory space should not be regarded as a panacea for communication problems, but rather as a technique for establishing a dialogue between users and designers at an early stage in the briefing and design process.

4.2 Classifying the Function of a Laboratory

It has been suggested that more fundamental than a classification of laboratory function by activity is one by patterns of education and communication. 'Scientific communities must be discovered by examining patterns of education and communication before asking which particular research problems engage each group.... Scientific knowledge is intrinsically a group product and neither its peculiar efficiency nor the manner in which it develops will be understood without reference to the special nature of the groups that produce it.'[7]

A system of analysing laboratory function introduced by Ruys[8] classifies practical science first by its *objectives*, then by *process complexity*, and finally as a *matrix* of these two categories. Ruys' classification of practical science by *objectives* is further divided into four groups: basic science, applied science, invention, and analysis. Conspicuously absent is the teaching laboratory. We should assume five categories as more appropriate therefore, and these can be summarized as follows:

i) *The basic science laboratory*, which undertakes fundamental discovery research with no immediately apparent application, usually at bench scale, and in a laboratory personally controlled by an individual research leader.

ii) *The applied science laboratory*, which produces useful results or verifications which can be applied to product development, usually with a larger team of workers, and relying on a greater use of benches, fume cupboards, and sometimes pilot-scale production facilities.

iii) *The invention laboratory* — often the 'sharp end' of pharmaceutical research — where marketable products are invented. The facilities in the laboratory may be similar to the applied science laboratory, but more people are involved due to the commercial pressures to accelerate product development.

iv) *The analysis laboratory*, which serves the other types of laboratory, and embraces a different type of activity: 'Analytical laboratory personnel usually make no judgements as to the value of the data they produce. Therefore, like basic science laboratories, these laboratories are concerned with accuracy, truth and specific knowledge.'[9] Analytical laboratories are difficult spaces to design, because they make intensive use of electronic equipment, and undertake often highly repetitive work.

v) *Teaching laboratories*, which can require lightly-serviced bench space for in excess of sixty students, plus a teaching area and ancillary areas for preparation and technicians. Safety design is of particular importance, as many of the occupants are unused to working in a laboratory environment with hazardous substances.

Ruys' classification by *process complexity* refers to single-project and to multiple-project laboratories, and within each of these classifications are single-procedure and multiple-procedure laboratories.

These ideas on the breakdown of laboratory function by assessing scientific objectives and process complexity are explained in more detail in Figures 2 and 3 respectively.

The resultant *matrix* developed by Ruys, and shown in Figure 4, forms a useful component in the early stages of gathering a brief, if only by asking the users to identify within the matrix the characteristics of their particular laboratory.

The benefit and clarity of the above system of classification can be seen when comparing to a more conventional classification system, such as that proposed by DiBerardinis,[11] which lists fifteen categories of space usage: general chemistry, analytical chemistry, high toxicity, pilot plant, physics, clean room, controlled environment, high pressure, biological safety, clinical, animal housing, teaching, radiation, anatomy/pathology, team research.

General Lab Descriptions by Scientific Objective	Basic	Applied	Invention	Analysis
Management control	One person very influential	One person with corporate influence	Corporate with personal influence	Corporate with personal influence
Project determinist	Scientific curiosity	Problem stater	Market needs	Customer
Output	Understanding and knowledge	Concepts and solutions to specific problems	Products	Data
Accountability	Scientific peers	Problem stater (military, society, corporate, etc)	Problem stater (market-place and profit)	Data customer
Value of results	Universal truth	It works under these conditions	How to make it	Accuracy, consistency, turnaround

Figure 2 *Laboratory classification based on scientific objectives, adapted from Ruys*[10]

General Lab Descriptions by Process Complexity. Issues:	Single-Project Labs		Multi-Project Labs	
	Single Procedure	Multiple Procedure	Single Procedure	Multiple Procedure
User control over project changes	User controlled	User controlled	More control by organization	Highly driven by organization's needs
Facility control by users	Usually controlled by user	Usually controlled by user	More control by organization	Driven by organization's needs
Rate of project changes	Mixed; >9 months typical	Mixed	Fast change; projects	Fastest; projects <6 months
Predictability of change	Predictable	Predictable, with some surprises	Hard to predict because of organizational change	Difficult to predict
Complexity of facility	Less complex	Somewhat complex	Somewhat complex	Very complex

Figure 3 *Laboratory classification based on process complexity, adapted from Ruys*[10]

SCIENTIFIC OBJECTIVES

	Basic	Applied	Invention	Analysis	Teaching
Process complexity					
Single project					
Single procedure					
Single project					
Multiple procedure					
Multiple project					
Single procedure					
Multiple project					
Multiple procedure					

Figure 4 *Laboratory classification matrix, developed from Ruys' simpler matrix*[10]

5 Brief-Gathering Techniques

'Brief-gathering determines requirements. Design meets those requirements'

The initial stages of laboratory design are sufficiently complex to warrant a structured approach to brief-gathering, and this invariably requires documentation in the guise of Room Data Sheets (RDSs). This ubiquitous approach has been criticized for gathering data which presuppose the designation of activities carried out in rooms. Hospital architects will, for example, use a standardized method of brief-gathering which gathers information on the *function* which is to be accommodated. It is then up to the architect to propose whether this function should occur in one or more rooms, or indeed not in a room at all.

An example common to both laboratories and hospitals is the office. The laboratory project's RDS might describe it as a single-person office of a certain standard, whereas the hospital project's RDS might refer to a requirement to accommodate someone requiring desk and meeting space, with privacy. The latter is inevitably a more time-consuming procedure, and often results in re-invention of the wheel, but is one which can be accommodated within the extended approval stages associated with most hospital design. Because there is less time available for elaborating laboratory projects, it will be assumed as inevitable at the outset that the particular requirement will be met as an office, and the design discussion will be concerned with the size, services, furniture, fittings and finishes of the office.

There are many types of RDS, and most institutions have developed their own. They all cover the same range of information, some in more detail than others. Not unusual in this respect is the computerized JohnData[12] system, the first sheet of which is shown in Figure 5.

The JohnData computer programme requests information under five sections:

- *General and Building*: Population and Time (occupancy); Relationships (to other rooms and buildings); Special Risks; Items to be Supplied and Stored; Items for Disposal; Items for Reprocessing; Doors; Windows; Ceilings; Walls; Floors; Insulation; Special Provisions.

- *Equipment and Furniture Equipment*: Loose Furniture; Special Equipment.

- *Mechanical Engineering Services:* Heating; Ventilation; Sanitary Fittings; Sinks; Drainage; Gases; Wet Services.

- *Electrical Engineering Services*: Power; Lighting; Communications; Computing; Special Provisions; Fire Services; Security.

- *Connections to Equipment*: Connections into equipment; Connections out from equipment.

ROOM DATA SYSTEM — GENERAL & BUILDINGS DATA   Date:

==   Page: 1

<ROOM IDENTIFICATION DETAILS>

Room No ....   Title ....

Discipline ....

Area ....   Status ....

Revision — No ....   Date ..../..../....

<GENERAL & BUILDINGS>

FUNCTIONS: ....

....

ACTIVITIES   ....

....

CRITICAL   ....
DIMENSIONS   ....

<POPULATION & TIME>

Normal — No ... Time .... Maximum — No ... Time ....

<RELATIONSHIPS>

(a) Separation — essential from   ....   desirable from   ....
(b) Proximity — essential to   ....   desirable to   ....

<SPECIAL RISKS>

Protect from: Noise/vibration/smell/damage/radiation/magnetic fields/dust/infection/...............

Liable to create: Noise/vibration/smell/radiation/magnetic fields/dust/infection/steam/heat/noxious
gases/ ....

Other risks: Fire/theft/................

<ITEMS TO BE SUPPLIED AND STORED>

Hardware/stationery/food/general stores/instruments/pharmacy/inflammable liquids/clothing/bedding/
cages/...............

<ITEMS FOR DISPOSAL>

Burnable ....   Non-burnable ....
Disposal by: Paper sack/truck/macerator/incinerator/other

Figure 5 *Introductory sheet to JohnData RDS system*

In the applications of any RDS system four principles should be observed:

i)   Data input should be coordinated to ensure that designers and clients do not enter information in parallel, and a single point of entry into the RDSs is available. JohnData's one master PC interactive disk secures this arrangement and can be entered and altered, allowing the design team to make initial entries, after which the client team can make their contribution. The programme will highlight at each handover any changes which have been made.

ii) An important element in the management of an RDS system is to identify at all stages items borne either by the project budget, or by a separate budget. Quantity surveyors will invariably base their cost on the content of the RDSs, and it is useful to distinguish those items sponsored by a separate budget.

iii) An RDS system should be managed to give the design team the information when they need it. It would be inappropriate, for example, to spend months collating data down to the number of plug sockets in each room, when the design team is more concerned with overall room numbers, room areas, relationships, and plant requirements. The solution is not to prepare preliminary data sheets, but to build up comprehensive RDSs in *passes*. The design team will aim in the *first pass* to establish outline and essential information, which can then be used for an initial design, sufficient to make a planning application and to determine project budgets within a 10% cost tolerance. While this design work is proceeding, the *second pass* can continue, leading to a *third pass*. The scope of each pass must be agreed between the client and the design team to suit the project's requirements.

iv) A formal method should be established for checking the evolving design against the RDSs. This also enables clients to see their requirements translated into a design and often results in redefined requirements.

Generalized procedures for data gathering should always be questioned. Individual methods are more suitable for particular organizations with their special circumstances. Nothing illustrates this better than the strategy adopted for refurbishment projects at the National Institutes of Health (NIH) in Washington, DC. The NIH has overcrowded buildings of poor quality, but relies on offering scientists facilities tailored exactly to their research requirements. Refurbishments of existing areas are often urgently needed. The NIH facilities manager, a bench scientist by background, builds scale models of the area to be refurbished, complete with furniture, fume hoods and scaled people. A typical model is shown in Figure 6. The occupant then agrees to the three-dimensional layout, the model is Polaroid-photographed and signed by both parties, and the NIH's construction team commences the work within days. This use of models, which will be referred to again later in Section 6, is the most dependable method short of full-scale mock-ups to ensure that the client's expectations coincide with his or her requirements.

Figure 6 *The working model used at the NIH, Washington DC for refurbishment planning*

6 Three Illustrated Case Studies

Three case studies illustrate different aspects of how a client's brief is met and achieved in a building. The first two concern the design of laboratory modules for Anglo-American pharmaceutical companies operating in the UK, and the third describes the design of a new laboratory building on a university campus.

6.1 The First Case Study

The client required a research module built to a common structural dimension which could accommodate equally effectively, and without any alteration in the overall dimensions, either chemistry laboratories, biology laboratories, or animal houses. The resulting module was to be applied to all new buildings in research sites in the UK, and possibly internationally.

A series of 'clinics' was established at which the designers met the user groups' representatives. Designs for the modules were developed at these meetings, and were illustrated with three-dimensional computer-aided drawings, models, and finally full-scale mock-ups.

User involvement was beneficial. A preference for a particular feature or spatial allocation led to its incorporation into the design. The module therefore evolved considerably, and the users suggested a range of features which were innovative within the company. These included:

- Write-up areas on the window side of the building;

- Direct access to the write-up areas without having to walk through the laboratory;

- An allocation of two 2.0 m fume cupboards per bench scientist;

- A separate storage area between two modules to prevent the accumulation of equipment within the working laboratory space;

- Proper ergonomic design of each write-up station to accommodate VDU terminals with filing and personal storage; and

- A view onto a landscaped area from each write-up station.

Figure 7 shows an early design for the module, while Figure 8[13] shows a model of the final design, which includes the features described above. The full scale mock-up of the chemistry module which, subject only to minor amendment, was approved by the user and client groups, is shown in Figure 9.[14]

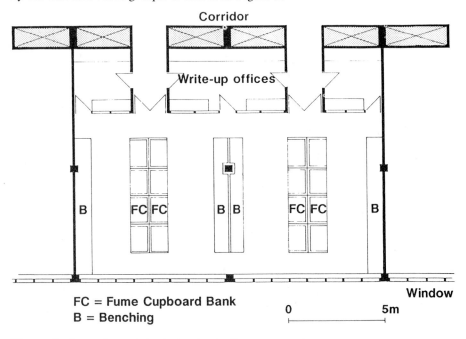

Figure 7 *An early module design in the first case study*

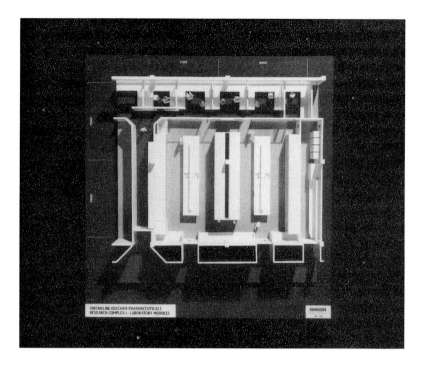

Figure 8 *The final design in the first case study*

Figure 9 *Built consultation mock-ups of the final design in the first case study*

6.2 The Second Case Study

This second study illustrates how, within the culture of an organization, preferences can change during a relatively short period.

The client commissioned a major multidisciplinary laboratory, which was completed in 1992. For the commission of a second laboratory building whose design commenced in 1992, a different approach to modular layout was taken. Though the client's representatives were essentially the same people, the emphasis for the second building was on unidisciplinary laboratory modules which could be used, with minor adaptation, for chemical or biological research.

This reflects perhaps a new dimension to design in private companies, where buildings and facilities need to adapt to changing research and development programmes in an increasingly competitive market.

Figure 10[15] shows the plan for the standardized chemistry module for the first building, designed in 1988 and completed in 1992. Of note are the 'outboard' write-up areas which can only be reached through the laboratories, and the relatively small (1.2 m) fume cupboards. The completed module is shown, ready for occupation, in Figure 11.[16] By contrast Figure 12[17] shows the proposal for the new general research module, designed in 1992. It is of interest to compare these two modules with those designed in the first case study. Both clients are large international pharmaceutical companies, and yet the culture within each company is quite different, particularly in the preferences of the users for the design of their laboratories.

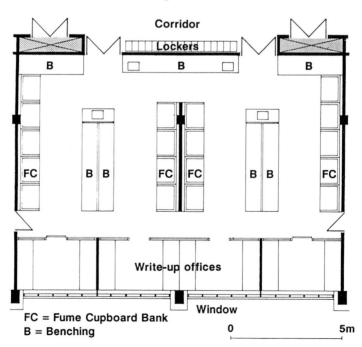

Figure 10 *The module designed in 1988 referred to in the second case study*

Figure 11 *The completed module from Figure 10, shown ready for occupation in 1992*

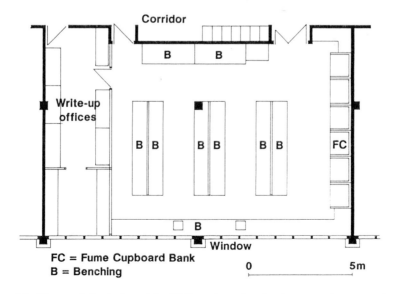

Figure 12 *The module designed in 1992, referred to in the second case study*

6.3 The Third Case Study

The new building for the Department of Soil Science at Reading University's Whiteknights campus is largely funded by the University Funding Council (UFC).[18] The UFC allocates an unrealistically low budget to research buildings, and the challenge in designing this particular building is first to produce one of attractive appearance, especially given its dominant and prominent location on the campus; secondly to provide the department with as much space and facilities as can be achieved within the budget; and thirdly to achieve a compromise acceptable to all parties between cost and function.

The identification of the client was particularly complex. Five separate client interests could be identified:

i)   The departmental teaching and research staff and students;

ii) The university's building office;

iii) The vice-chancellor and bursar;

iv) The UFC (perhaps the most influential);

v)   The local planning authority, representing the public interest (perhaps the most disruptive).

Whilst pharmaceutical companies can allow operational choices to overrule cost-led design, few such concessions can be permitted in a UFC-funded building. The challenges in determining and meeting the clients' requirements are quite different. Moving from another campus where the department occupied a number of substandard buildings, the users often had to decide on those facilities they could manage without. The strategy has been to make allowance in the building fabric wherever possible for the subsequent addition of, for example, extra fume cupboards. The main preoccupation during briefing was to determine affordable features and to prioritize choices.

Meeting the clients' requirements in this case has therefore been an exercise in design compromise, or 'value designing', to borrow a phrase from value engineers. Figure 13 shows a simplified section through the building, while a typical floor plan is shown in Figure 14.

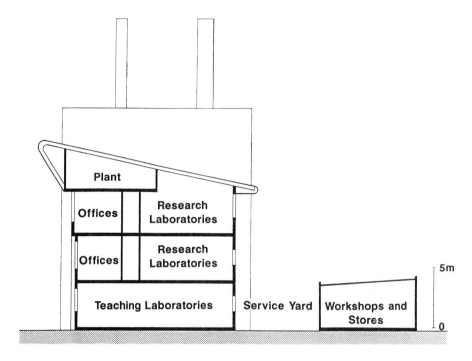

Figure 13 *A section through the building for Reading University referred to in the third case study*

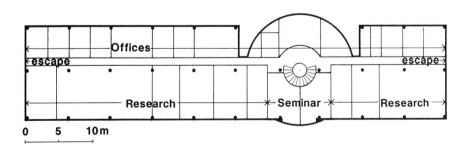

Figure 14 *A typical floor plan of the building in the third case study shown in section in Figure 13*

The objectives of designing within such tight budgetary constraints for this third case study have been:

- To divide the building into economic zones, and to group together highly serviced and therefore expensive spaces, and correspondingly also to group together the cheaper spaces;

- To remove from the building the cheapest spaces such as stores, workshops and soil holding facilities, and to house them in outbuildings constructed to a standard similar to domestic garages;

- To site on the ground floor the two large teaching laboratories for undergraduates, thereby reducing considerably the number of people who would need to use the upper floors;

- To limit the palette of materials both internally and externally to a small range of durable and good quality products;

- To design the building for a shorter construction period compared with other facilities by incorporating a simple repetitive structural frame, using repetitive window sizes where possible, avoiding wet trades such as plastering, eliminating the traditional suspended ceilings, simplifying and shortening service routes, and by positioning the building on the site so as to allow clear access on all sides for construction.

The building, shown in Figure 15, will be completed and occupied in 1994.

These three case studies have demonstrated different aspects of laboratory design, and the relationship between the requirements and gathering of the brief, and its transformation into a built design.

There are numerous examples of major laboratory buildings which have either succeeded or failed to meet their users' expectations. Buildings which might be considered successful include the NIEHS Laboratories at Research Triangle Park, USA, the NIH at Washington, DC, the new Pfizer Laboratory at Sandwich, Kent, UK,[3] ICI's Laboratories at Jeallots Hill, Berkshire, UK,[19] and the Laboratory of the Government Chemist at Teddington, UK.[20] Buildings whose design do not perhaps reflect entirely the requirements of their occupiers include Bristol Myers Squibb at Wallingford, Connecticut, USA, and Burroughs Wellcome at Research Triangle Park, North Carolina, USA.

Our experience in laboratory design leads to one dominant conclusion — applicable to large, small, new or refurbished projects — that the widest possible consultation with, and involvement of, the building's future occupants at the briefing stage will, perhaps more than anything else, contribute to an effective and valued building.

Figure 15 *The new building for the Department of Soil Science at Reading University, to be completed in 1994 (the third case study)*

7 Footnotes and References

1. Briefing is known in the USA and Canada as programming.

2. 'Royal Institute of British Architects Guide to the Standard Form of Agreement', RIBA, London, 1992.

3. Building 503 for Pfizer Central Research at Sandwich, Kent, designed 1988-1990, completed 1992. Johnson Partnership Limited, Architects, Engineers, and Quantity Surveyors. Colin Bloch and Thomas Gibbard, Architects.

4. C. Bloch, 'Procuring Laboratory Construction Contracts: Guidance for Chemists', in *Analytical Proceedings*, 1992, **29**, 503.

5. T. Ruys, 'Handbook of Facilities Planning; Volume 1: Laboratory Facilities', Van Nostrand Reinhold, New York, 1990.

6. Reference 5, p. 1.

7. T.S. Kuhn, 'The Essential Tension', University of Chicago Press, Chicago, 1989, quoted in reference 5, p. 2.

8. Reference 5, pp. 4-11.

9. Reference 5, p. 5.

10. Figures 2 and 3 are based on reference 5, pp. 5-6, but have been adapted to include teaching laboratories in Figure 3.

11. L.J. DiBerardinis, J.S. Baum, and M.W. First, 'Guidelines for Laboratory Design: Health and Safety Considerations', J. Wiley, New York, 1987, pp. 22-23. (2nd Edition, by L.J. DiBerardinis, J.S. Baum, M.W. First, G.T. Gatwood, E. Groden, and A.K. Seth, J. Wiley, New York, 1993).

12. JohnData was developed for Johnson Partnership Limited in conjunction with Pfizer Central Research, Copyright Johnson Partnership Limited.

13. Photographed by Alison Needler, copyright Johnson Partnership Limited.

14. Photographed by Alison Needler, copyright Johnson Partnership Limited.

15. Copyright Johnson Partnership Limited.

16. Photographed by Alison Needler, copyright Johnson Partnership Limited.

17. Copyright Johnson Partnership Limited.

18. Johnson Partnership Limited, Architects, Engineers and Surveyors, Colin Bloch, Architect.

19. IDC, Architects and Engineers, Stratford-on-Avon, Warwickshire.

20. PSA, Architects and Engineers, London. Gordon Wilson, Architect.

CHAPTER 13

Design of the Rhône-Poulenc Rorer Central Research Laboratories at Dagenham

J. SALMON

1 Introduction

1.1 Research at Dagenham

In 1934, May and Baker, a fore-runner of Rhône-Poulenc Rorer (RPR) in the UK, transferred its pharmaceutical business from Battersea, London to a green-field site on the eastern outskirts of London at Dagenham. The company wished to take advantage of the large pool of labour available in the area for packaging operations, following the establishment of the Ford Motor Factory at its riverside works.

A research organization was established at the Dagenham site and much fundamental research was undertaken in the new facilities. The highlight of this period was the discovery of M&B 693, a pyridinyl sulfonamide which was to be the first of a new generation of anti-bacterial drugs, and a major step forward in the early period of medicinal chemistry. In the early 1950s, plans were made to build a completely new Research Institute on the Dagenham site. Construction began in the mid fifties and the synthetic chemistry building (D41) was completed in 1960.

1.2 Impetus for Change

By the mid 1980s it had become clear that the synthetic chemistry building, D41, was unable to support the standard of synthetic chemistry necessary for the development of new medicinal products. In particular the number of fume cupboards, their location and performance failed to meet current standards for an acceptable laboratory environment.

There had also been a number of significant changes within the pharmaceutical industry which were affecting the environment for competitive research:

- *Increased resources*
  There had been a trend towards the establishment of large multinational pharmaceutical companies to generate the profits required to fund effective research. These companies had usually been formed from the takeover and merger of existing pharmaceutical organizations.

- *Safety of medicines*
  Following the harrowing events after the introduction of thalidomide, governments of most developed countries introduced legislation for the safety testing of new candidate drugs. This has led to longer drug development times, increased costs, and the pharmaceutical industry has had to review its research procedures and attempt to focus its efforts to ensure that cost effective research is undertaken.

- *Advances in chemistry*
  Significant developments in organic and organo-metallic chemistry occurred during the period from 1960 to 1985. New and powerful reagents were introduced which have enabled chemists to make dramatic progress in chemical synthesis. However, the reactive potential of these reagents is also reflected in their toxicity.

- *Health safety legislation*
  A greater awareness of the possible risks associated with handling chemicals has resulted in legislation to protect employees, for example the Health and Safety at Work *etc.* Act 1974, and the Control of Substances Hazardous to Health (COSHH) Regulations, 1988. A significant development for laboratory safety was the Draft Document on the installation and operation of fume cupboards (DD80) issued by the British Standards Institution and published in a slightly modified form as British Standard (BS) 7258.[1]

- *Focused research*
  It was realized in the early 1980s that pharmaceutical research had to be focused if its objectives were to be met effectively. Traditionally, research groups had worked in diverse areas, often with little chance of success. But by channelling efforts into fewer projects and providing the necessary financial and technical resources the opportunities for discovering new products could be improved. It was also recognized that many new compounds were being lost due to delays in their development. It was essential therefore to reduce the time span for development, and to adopt a ruthless approach to the fate of any compounds with development problems.

1.3 Planning for the Future

In the mid 1980s a study was undertaken in conjunction with a consultancy company[2] to determine if a refurbishment of our existing facility was possible. Building D41 had been considered an advanced design when constructed in 1960, but increased knowledge and changing legislation had rendered the facilities obsolete. The main problems included the number and location of fume cupboards (adjacent to doors), the poor performance of these fume cupboards for synthetic chemistry (due to an inadequate air supply to laboratories), and the predominance of small laboratories in which just one person might be working ('lone working').

The consultancy company devised a scheme to upgrade D41 floor by floor, thus enabling research to continue unhindered throughout this period. This would have allowed new fume cupboards to be installed to DD80 requirements. To ensure that the fume cupboards did perform to the required standard, it would have been necessary to provide supply air through new duct-work fixed outside the building (due to limited space within the frame of the building).

The consultants' final report suggested that refurbishment of D41 was feasible with minimum disruption of current research effort. The cost would have been £1-2 million less than plans for a new research facility.

This report was considered in detail, and it was decided that although refurbishment was feasible, the alternative, to build a new facility, had several advantages. For example, although the total cost was higher than renovating D41, it would be possible to operate to the highest safety standards in the new building in compliance with all health and safety legislation, and D41 would still be available if needed for less demanding activities. An important factor taken into consideration was the real effect on the continuing research effort caused by the proposed floor-to-floor refurbishment. The general appearance of the Dagenham Research Centre would also be preserved by not having to modify D41 externally with ducting.

Plans to refurbish D41 were, therefore, abandoned, and the decision was taken to design and build a new research facility. This decision was also in agreement with the philosophy of the research management at Dagenham that the best facilities should be provided which could be justified financially, and that all facilities both old and new should be operated and maintained to the highest possible standards.

By making these changes the research effort at Dagenham could become fully focused and the potential for drug discovery optimized.

2 The 'Design and Build' Contract — The Rhône-Poulenc Rorer Project Team

Company policy for large projects was to use a 'Design and Build' contract and this was the basis for the construction of the new facility. The building to be constructed on site number 38 at the Dagenham Research Centre became known as the New Chemical Research Laboratories, D38.

A recently established practice at the company was to develop user-led teams when undertaking projects of this complexity. Past experience had demonstrated that it was essential that user representatives should play a central management role in the creation of the specification, the design development, construction and commissioning procedures. It was also important that the user representatives were sufficiently knowledgeable to decide issues at contract meetings and not have to refer questions back to senior management.

Experience had also shown with highly technical complex projects that the specification should be drawn up with a degree of foresight to allow for future developments. Given a free choice many scientists might be tempted to specify a new building which was merely an improvement on their existing laboratories, rather than a 'quantum leap' forward in imagination. User representatives need to question all existing procedures and consider new alternative practices if an efficient and successful research facility is to be built.

User involvement in design would ensure that the users' specification was faithfully reproduced in the new facility. Major problems can occur with a completed facility because accumulated but relatively minor problems have been resolved without understanding fully the consequences for laboratory experimental work.

With a highly-serviced complex building, the commissioning engineer must have a high profile role during the design, development, construction, and the final commissioning period. It is essential that the commissioning period is adequate so that both basic and complex scientifically-based commissioning can be completed. The users' scientific expertise must be enlisted to ensure that all possible hazards are examined during the commissioning of the facility. For example, will the ventilation and drainage systems contain all odorous and toxic fumes and prevent their dispersal into the air-input system despite the vagaries of English weather?

The user-team for the construction of D38 is shown in Figure 1. The Director of the Dagenham Research Centre[3] was the project leader, but the day-to-day management of the project was undertaken by the Project Control Manager.[4] All meetings with the building contractors were attended by the Project Engineering Manager's team, the Project Control Manager, the Contracts Consultant and the Senior User Representative. Apart from my role as Senior User Representative, I also hold the position of Chemical Safety Adviser for the Dagenham Research Centre, and my involvement ensured that the many detailed decisions were made only after considering safety aspects of the design. The high safety profile was to be significant in drafting the users' specification, in developing the design, in prototype testing, and in the final facility commissioning. Safety considerations applied particularly to laboratory layout and to fume cupboard, ventilated cold room, chemical lift, and solvent store design.

As Senior User Representative, I was also responsible to a user-group[5] whose main tasks were to provide information for the compilation of the users' specification, to keep staff informed of progress, and to address any major problems that I could not immediately answer at the contract meetings.

3 Selection of a Contractor

The company had employed 'Design and Build' contracts for all major projects in the UK. The advantages of this type of contract are that costs can be tightly controlled and the contractor can work to a comprehensive user specification.

The selection of the main contractor to undertake the construction of the New Chemical Research Laboratories D38 was crucial for the success of the project. Most contractors have limited experience in the design and construction of such facilities and are unlikely to understand the differences between analytical chemistry, process chemistry, physical chemistry or synthetic organic chemistry laboratories.

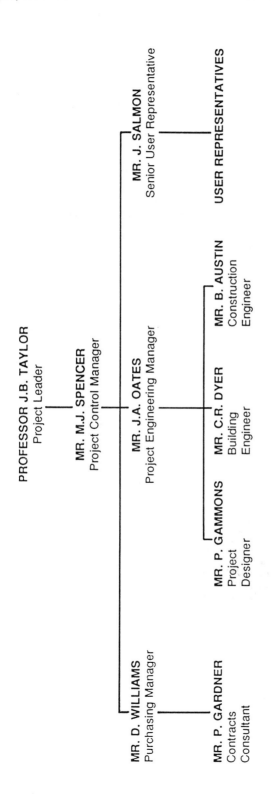

Lines of Responsibility

Figure 1 *The Rhône-Poulenc Rorer project team for the design and construction of D38*

For the project to achieve its primary aims, it was essential that the consultants used by the contractor had a flexible working relationship with the Rhône-Poulenc Rorer project team. The contractor had to interpret the user specification whilst taking into account the in-house operating philosophy and culture. It was important that design problems were resolved amicably at the design stage. A high level of respect for the particular expertise of the individual members of the RPR project team had to be established at an early phase in order to reduce the scope for argument and confrontation, and ensure the completion of a quality building.

The selection of a contractor for D38 began with a letter to sixteen companies asking if they would be interested in constructing this new facility. These companies had been chosen from organizations we had used previously and from those contractors with a proven track record in laboratory design and construction.

By the closing date given for a reply (one week), twelve companies had expressed interest, one company said they were too busy to tender and three did not reply by the closing date. The Rhône-Poulenc Rorer project team then drew up a second list of eight companies from the twelve positive replies. A representative group from the Rhône-Poulenc Rorer project team then interviewed each of the eight companies at their own premises. The contractor was asked to assemble the entire team who might be involved in the project: architectural, structural, mechanical, project, and site management staff.

Interviews generally lasted 3-4 h and involved the Project Engineering Manager, the Contracts Consultant, the Project Designer and the Senior User Representative. The interviews were searching in nature and the attitude of staff under pressure was noted. There were some interesting revelations — for example, a mechanical engineering contractor (a major British company) whose representatives admitted that they were not sure what a fume cupboard was. This hardly implied confidence in the ability of that company to construct fume cupboards to the desired operating standard.

The wealth of information gained from this exercise enabled the Rhône-Poulenc Rorer project team to select four finalists from the original sixteen companies. It was decided that these four companies should be allowed one month to produce a concept design for the proposed facility. As an inducement, Rhône-Poulenc Rorer agreed to pay a nominal fee. Payments went up to £15,000, and the final designs became the property of Rhône-Poulenc Rorer.

Each contractor was allowed a single visit to the Dagenham Research Centre to meet the project team and discuss the specifications drawn up by Rhone-Poulenc Rorer. They were quoted a maximum acceptable price on which to base their designs. After four weeks each contractor was invited to present their proposed design to the Rhône-Poulenc Rorer project team at Dagenham:

Design Proposal A
Architecturally this was an interesting design, demonstrating the input of a dynamic architect in the team. The layout of the mechanical services to the laboratories was *via* the corridor roof space, and various questions were raised by the Rhône-Poulenc Rorer project team regarding the compliance of this design with fire and safety legislation.

The proposal was within the price range requested, but it had been noted at the previous interview that this contractor took a very rigid and inflexible attitude towards a 'Design and Build' contract. The Rhône-Poulenc Rorer project team thought that an increase in the total cost would be inevitable during design development.

Design Proposal B

The overall design of the building was reminiscent of a 1960's hospital and lacked an exciting architectural theme. The laboratories were very large, with a traditional layout of fume cupboards and associated services. We were aware, however, that the standard of the mechanical engineering was very high, perhaps even luxurious! Though the mechanical engineering aspect was a strong point, the laboratory design did not make the best use of available space.

The presentation was comprehensive, with a large number of specialists present. However, the proposal was approximately £3 million in excess of the specified maximum cost.

Design Proposal C

This design was the most exciting of the four architecturally, reflecting the influence of a strong and independent architect. An atrium was part of the design and the artistic presentation claimed this was a significant feature, but further examination of the plans suggested something less. There was concern within the Rhône-Poulenc Rorer project team that valuable ground space would be lost at significant cost. Overall the design proposal was acceptable, and remained within the maximum cost.

Design Proposal D

This design proposal again featured an atrium as a feature of the design. Unlike the previous proposal, this design used the atrium for a more practical purpose — to let in light to the centre of the building. Again, laboratory space would have been reduced though the design proposal was generally acceptable, and remained within the required maximum cost.

These four design proposals were considered by the Rhône-Poulenc Rorer project team together with information obtained from interviewing the contractors. Two design proposals were discarded immediately. Proposal B was considered to be over-engineered, expensive, and lacking in inventive design. Proposal A was rejected for safety and fire regulation reasons and an apparent inability to absorb fully the users' preliminary specification.

The remaining proposals, C and D, were both considered to have equal merit. Proposal C was more architecturally impressive, but both buildings were suitable for the basis of a full design. Each of these plans featured an atrium, and the Rhône-

Poulenc Rorer project team were concerned about using space for this purpose, especially as the total area of land available was restricted.

The project team asked each company to reconsider their designs for a further week. In particular they were asked to consider the necessity for an atrium. The two contractors then made a second presentation to the Rhône-Poulenc Rorer project team.

An impressive plea was made by the architect to justify the presence of the atrium in design proposal C. Some minor changes reflecting the project team's comments had been made, but the contractor considered that the atrium was an essential ingredient in the overall concept of the building.

The second contractor had accepted comments regarding design proposal D. The layout of the building had been reconsidered with the removal of the projected atrium. These changes fully reflected the concerns of the Rhône-Poulenc Rorer project team.

After further consideration, it was decided to offer the contract to the company[6] which had supplied design proposal D. The Rhône-Poulenc Rorer project team were impressed with this company's constructive and flexible attitude to overcoming design problems within the confines of a 'Design and Build' contract.

4 Specifications for Building D38

A number of basic requirements for inclusion in the building specification were established by the Senior Research Management and the project team from visits to modern research laboratories operated by our competitors in the UK, USA, and France. These included:

i)   The size of building suitable for 65-70 working chemists;

ii) A laboratory area designed for up to eight working chemists;

iii) All synthetic chemistry to be performed in fume cupboards operating to the accepted draft British Standard available at that time (DD80);

iv) A clean office environment, physically separated from the laboratory area;

v)   The systematic writing-up of notebooks to be actively discouraged in the laboratory area;

vi) The design of the laboratories to take into account current and foreseeable safety legislation.

Our visits to other modern research laboratories were instructive. We were very impressed by many of the facilities we saw, for example, the design of the maintenance ducts at the Wellcome Laboratories at Beckenham and the quality and flexibility of the plant installed at the Laboratory of the Government Chemist at Teddington.

Users' requirements were collated by the Senior User Representative and the user-group whose members (scientific and technical staff) were selected for their ability to acquire and disseminate information, and to take an active part in the decision making processes. These far-ranging and difficult decisions needed perception and imagination for they would influence working practices at Dagenham well into the next century.

Each section of the Discovery Chemistry Department was asked to provide a list of essential and desirable features for the layout of the laboratory, fume cupboard design, number and position of services and design of specialist elements (*e.g.* chemical lift, solvent store, and a ventilated cold room). This exercise, completed within 2-3 weeks, was followed by a short but intensive consultation period.

Though it was company policy to use by-pass fume cupboards with epoxy resin liners for synthetic chemistry, all other aspects of the design evolved from the users' requirements. Our aim was to provide all *essential* features and as many other *desirable* elements as could be financially justified.

5 The Design and Construction of D38

5.1 Company Policy

Building D38 is the first of a new generation of laboratories to be constructed at Dagenham. With this philosophy in mind, the company decided there was no need for D38 to harmonize with the other existing but ageing facilities on the site. It was to be designed and built to meet the highest safety and environmental standards using the most modern materials that could be bought with the financial resources available.

5.2 The Layout of D38

The New Chemical Research Laboratories Building (D38) is a post-stressed cast *in situ* concrete construction, clad with epoxy-coated mild steel panels (Figure 2). The north-facing front of the building is totally double-glazed, and features the main entrance and staircase, and the passenger lift. The lift motor room on the fourth floor enhances the visual interest of this part of the building, and the roof line is broken by a penthouse meeting room on the third floor.

The southern aspect of D38 is dominated by an impressive 40 metre ventilation exhaust chimney (Figure 3) consisting of three exhaust ducts, supported by a circular steel frame from ground level. These ducts emerge from the building at the third floor plant room level. The services' gantry enters the building at the second floor level, and provides steam, water, compressed air and bulk nitrogen supplies generated elsewhere on the site. Because D38 is built on a sloping site, the rear service entrance of the building is elevated in relation to the service roadway. Ramps have been provided to the entrance, and chemicals can be transported to and from this entrance with the aid of a purpose-built electric vehicle (Figure 4).

Figure 2 *The New Chemical Research Laboratories Building (D38)*

Figure 3 *The main exhaust chimney for D38*

Figure 4 *Specialist electric handling vehicle*

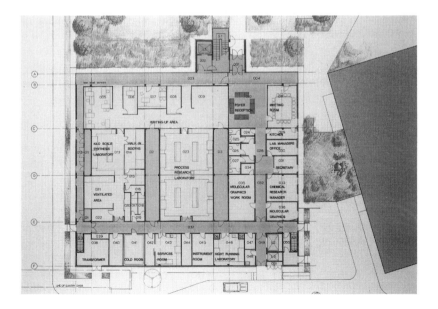

Figure 5 *Ground floor plan for D38*

A detailed plan of the ground floor is shown in Figure 5. The foyer and reception area plus associated meeting room, kitchen and toilet facilities (including an invalid toilet) are not under Cardkey security control, and staff from other parts of the site, who may not have permission to visit the high security, potentially hazardous, areas, are allowed to use this meeting room and its associated facilities. The east side of the ground floor has management offices, and the computer-aided drug design facility which is so essential for modern drug discovery. In the middle of the ground floor there are two laboratories, a Process Chemistry Laboratory (MO23) and a Kilo Scale Synthesis Laboratory (MO13). Office and service areas for these two laboratories are also on the ground floor. All functional laboratory facilities are separated by maintenance service ducts.

The plan of both the first and second floors is shown in Figure 6. There are three Discovery Chemistry Laboratories situated on each floor (MX13, MX14, MX15) with associated office and writing-up areas to the north and a service area to the south, respectively.

The general layout of the third floor is shown in Figure 7. There are three centrally positioned plant rooms immediately above their respective laboratory areas. The central maintenance ducts on the lower floors enter this level adjacent to the plant rooms. Above the general service area to the south of the building is an open plant area next to the main exhaust chimney. The northerly part of this floor is mainly open roof space plus the passenger lift shaft and the penthouse meeting room referred to earlier.

It can be seen from the three floor plans that D38 can be divided into three vertical segments. These laboratory segments are separated by the maintenance service ducts situated between each laboratory module. The maintenance service ducts have a one hour fire protection rating, and from a fire regulations viewpoint they can be considered an external part of the building (Figure 8).

5.3 Ventilation of D38

5.3.1 Ventilation Philosophy

Air movements and pressure differentials within D38 are designed and engineered to minimize exposure to chemicals in all parts of the buildings. A high degree of safety is built into the system to ensure that it can only fail in a safe mode.

Thus the ventilation system maintains a negative pressure in the laboratories in relation to the office, writing-up and services areas, and prevents hazardous fumes from entering the office and writing-up areas. Similarly, if a release of a potentially hazardous or flammable material occurs in the service area, ventilation is achieved by opening the laboratory door between the two areas, and allowing the fume cupboard extraction system to cleanse the affected room. The office, writing up and service areas are maintained at a slightly positive pressure relative to atmospheric pressure.

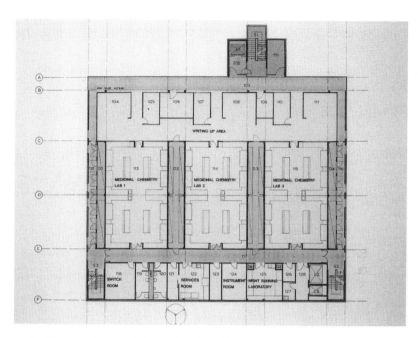

Figure 6 *First and second floor plan for D38*

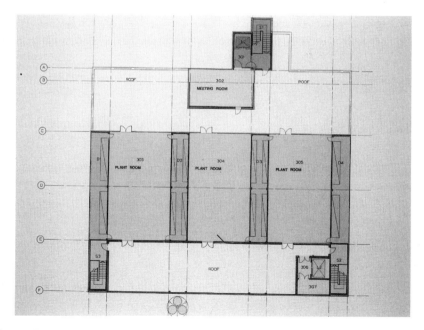

Figure 7 *Third floor plan for D38*

Figure 8 *Maintenance service duct in D38*

Figure 9 *Third floor plant room — primary fans and collection duct in D38*

5.3.2 Ventilation Design

The air-input system for D38 consists of individual air-handling units (situated in the third floor plant room) for each laboratory, the ground floor management area and the service zones. Air is drawn from the third floor level both to the north and south of the central plant room area. The air is tempered by passing it through hot water heater batteries; hot water (90 °C) is produced in the second floor plant room (M218) from steam generated on site. The air handling plant was designed to be able to heat the laboratories to a maximum temperature of 23 °C when the external ambient temperature fell to a minimum of -5 °C. For lower external temperatures it will be necessary to increase the water temperature above 90 °C to maintain temperatures of 23 °C and above in the laboratories.

The tempered air is then passed *via* insulated galvanized ductwork from the plant room, through individual maintenance service ducts to a ceiling plenum above each laboratory. Laboratory ceilings consist of Burgess ceiling tiles.* Air is generally fed into a laboratory through a band of 'live' ceiling tiles situated between the island benches and the fume cupboards. The safety screen on the island benches assists the directional air flow from the ceiling. Careful attention was paid to the design of the 'live' panel positions when prototype fume cupboards were tested in the manufacturer's test facility. In order to reduce any noise and draughts from the air input, the 'live' tile layout was developed to create a maximum air velocity of 2 metres/sec. Additional 'live' general ventilation tiles are situated above the central aisle of each laboratory and over the ventilated sink unit. A single air-handling unit supplies tempered air to the service areas on all three floors.

Though some air is extracted from offices, writing-up areas, store and service rooms, most of the air in D38 is extracted *via* the laboratory fume cupboards. Each laboratory contains eight fume cupboards operating to DD80. A single primary fan draws air from a pair of adjacent fume cupboards and a balancing damper is included in the system to ensure equal performance from each fume cupboard.

Air is extracted from the pair of fume cupboards *via* a PVC duct which is protected from fire by a glass wool/epoxy resin coating. The extracted air passes from the rear of the fume cupboards into the maintenance service duct, and from there into the plant room on the third floor level (Figure 8). The exhausted air is drawn by the primary fan and ejected into a horizontal collection duct situated above the fan unit shown in Figure 9. Each of the three plant rooms has two collection ducts. These collection ducts have an open louvre at the north end and are combined at the south end of the plant room *via* a sound attenuator before joining the main secondary fan. The main secondary fan draws air from the combined collection ducts and discharges it up one of the three main chimney ducts (Figure 10). Up to this point, the entire ventilation system is maintained at a negative pressure thus ensuring that no leaks of potentially toxic material can occur.

*A Burgess ceiling tile has a number of holes in it to allow air to pass through. Air movement is controlled by placing a plastic bag containing glass fibre or mineral wool on top of the ceiling tile in the ceiling plenum. If the bag is located directly over the tile, no air can pass through and the tile is 'dead', but if the bag is removed air can pass through and the tile is 'live'.

The pressure differential is maintained by ensuring that the air input to each laboratory is adjusted to provide approximately 95% of the extracted volume. The additional volume of air is drawn from the office, writing-up and service areas. The whole ventilation system is alarmed so, for example, if one of the primary extract fans fails, an alarm sounds and the laboratory air input is reduced to 50% of its normal volume to ensure that the laboratory is not pressurized.

Out of normal working hours, the ventilation system is reduced to 50% of its normal volume in the general laboratory areas. No overnight experiments are undertaken in these areas, and a separate laboratory on each floor has been established for running experiments overnight. The fume cupboards in the overnight laboratories have individual primary extract fans located on the southern open roof area (Figure 11). This arrangement prevents damage to the plant rooms if a fire occurs during unattended operations.

None of the air supplied to the main laboratories is recirculated. The rate of ventilation in these laboratories is dictated by fume cupboard design and DD80 operating standards. The nominal rate is approximately 50 air changes per hour.

The office and writing-up areas are mechanically ventilated with approximately 12 air changes per hour and partial recirculation of spent air. Spare cooling capacity from the computer-aided drug design facility is used to provide some comfort cooling in the office/writing-up areas during the summer months.

5.4 Office and Writing-up Areas

It is company policy to keep laboratory, office and writing-up areas entirely separate; thus laboratory staff do not have to undertake clerical activities in a potentially hazardous laboratory environment. In the old chemical research laboratories (D41) 'lone working' had also been a major concern. The combination of small laboratories, isolated offices, and extended working hours due to 'flexitime' working had led to restrictions being placed upon staff to ensure safety.

It was realized in planning the new building, D38, that although the office and writing-up areas had to be separated from laboratories, they needed to be adjacent with a good visual access between them. These requirements were put into practice by separating the laboratory and write-up areas with glazed partition units fitted with Georgian wired glass to maintain high safety standards (Figure 12). The positive air pressure maintains the writing-up area as a chemically clean zone. All chemicals are banned, and smoking, laboratory coats, gloves, and potentially contaminated items are prohibited. By keeping this zone free from chemical contamination, staff are allowed to consume refreshments at their desks rather than in remotely-sited tea rooms.

The office and writing-up areas adjacent to each laboratory incorporate an open communal space for laboratory staff and a separate office for the Section Manager. This office is constructed from epoxy-coated mild steel partition units which can be easily modified if necessary. A glazed wall partition allows a clear view from the office to the laboratory and write-up area. The write-up area itself is designed to be sufficiently large for all staff to have their own desk, chair and 2 metres of

Figure 10 *Third floor open plant area: collection ducts, sound attenuators and secondary main extract fans in D38*

Figure 11 *Third floor open plant area primary fans for night running fume cupboards in D38*

Figure 12 *View of a write-up area and laboratory in D38*

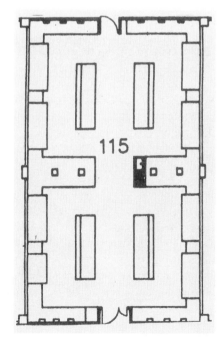

Figure 13 *A standard laboratory layout in D38*

bookshelves. It also contains coat hangers, space for filing cabinets, whiteboards, computer facilities, and a free open space for staff to meet and exchange ideas.

The communal writing-up area for laboratory staff is designed to allow staff to arrange their preferred layout of desks and associated equipment without obscuring their view of the laboratory. The result is that each write-up area has a different layout and character reflecting the occupants' predilections.

The office and writing-up areas for each laboratory are basically similar, with only small differences due to the geometry of the building. Other minor differences between floors are an additional office and library room on the second floor (M209), and a central computer network room (M109) in the equivalent position on the first floor.

The Department Manager on each floor has an office in a central position on the north side of the office/writing-up areas. These north-facing offices afford privacy for sensitive or confidential discussion. The first and second floor office and write-up areas are served by two small vending rooms adjacent to the main staircase which have been fitted with food and drinks machines. The offices and writing-up areas are generally carpeted, and are fitted out to a high standard.

5.5 Laboratory Layout

5.5.1 Standard Laboratories

The laboratory layout (Figure 13) provides a high standard of safety for both laboratory and maintenance staff. Maintenance service ducts between the laboratories mean that maintenance staff can work outside a potentially hazardous laboratory environment.

Though there are no external windows in the laboratories, lighting standards are good and a large amount of 'borrowed' light is available from the fully glazed front aspect *via* the office and writing-up areas (Figure 14).

Each laboratory is fitted with eight fume cupboards. The Discovery Chemistry Laboratories on the first and second floors contain seven standard fume cupboards and one column fume cupboard, whilst the Process Chemistry laboratory on the ground floor has six standard fume cupboards and two column fume cupboards. The fume cupboards are arranged in pairs with a section of ventilated bench between them. This ventilated bench area is meant for the use of balances or electrically-heated drying ovens. Ventilated cupboards for the storage of toxic reagents are situated below the fume cupboards. The four cupboards beneath each pair of fume cupboards draw 2.5% of the total volume of air extracted by each extraction system.

Each laboratory in D38 incorporates a wide central area running between the door to the office and writing-up area and the door to the service areas. Staff are encouraged to use this central area and not to walk immediately behind staff working at a fume cupboard, as this is poor safety practice and could affect the fume cupboard's performance. Approximately 1.5 m of bench surface are provided for each fume cupboard at a bench height of 900 cm (a higher bench might have restricted the

Figure 14 *General view of a standard laboratory in D38*

Figure 15 *Laboratory ventilated sink in D38*

view between laboratory and write-up areas). Wall benches are fitted with power sockets, cold water, and vacuum and compressed air lines.

The laboratory floor slabs of D38 are cast in one piece with no perforations. This prevents water leaking from one laboratory to another — one of the main flaws identified in the old building, D41. 'Island' benches in each laboratory are provided with power and vacuum services, but no water, as no route is available for the drainage. The only wet service in the ceiling voids of laboratories is the water supply from the maintenance service duct to an emergency shower unit placed near the service area door.

'Island' benches are fitted with toughened glass screens (to protect staff from any mishap in an adjacent fume cupboard). The user-group decided that drawer and cupboard units would be fitted under *all* laboratory benches in a ratio of 3:7.

Laboratory glassware is washed in Lancer washing machines in the service area on each floor, though some delicate items are washed and rinsed with acetone in the laboratories.

Acetone rinsing is carried out in a single large ventilated sink and drainer unit situated at the end of the peninsular bench (Figure 15). The ventilated sink is designed to reduce exposure to acetone vapour. Air, from a 'live' ceiling tile immediately above the sink unit, is drawn through a slot in the side of the sink unit. The drainer unit adjacent to the sink is ventilated through the base unit to ensure that any acetone dripping from vessels on the rack above is safely removed. The ventilation from the sink and drainer unit is exhausted *via* the maintenance service duct and the main exhaust chimney.

Three cupboards below the ventilated sink and drainer unit are also ventilated, and are used as waste disposal storage units. Company policy is to keep a minimum amount of solvent in each laboratory. A single small flammable solvent cupboard in each laboratory is used for storage of waste flammable solvent prior to its disposal, and a specialist solvent store is located on each floor in the service area adjacent to the laboratories (see 7.4).

The power ring mains are individually tripped, with the reset buttons situated in the service corridor. The fume cupboard switched socket outlets are individually tripped for safety reasons. Emergency power stop buttons are provided at each end of the laboratory, to shut down the power ring mains only. Lighting and fume cupboard extraction fans are not affected.

5.5.2 Kilo Scale Synthesis Laboratory

The Kilo Scale Synthesis Laboratory layout is shown in Figure 16. This laboratory produces the larger quantities of compounds required for further testing and development. The original specification was for the synthesis of kilogram scale quantities of materials using vessels of up to a maximum size of 50 litres.

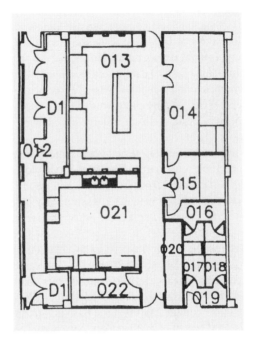

Figure 16 *Kilo Scale Synthesis Laboratory layout in D38*

Figure 17 *Booth area in the Kilo Scale Synthesis Laboratory*

Figure 18 *Booth doors in the Kilo Scale Synthesis Laboratory*

The Kilo Scale Synthesis Laboratory has two standard fume cupboards identical to those provided in the standard laboratory, and a column fume cupboard adapted to zone 1* electrical requirements.

After considerable debate it was decided to design and install in the Kilo Scale Synthesis Laboratory a number of specialist booths rather than the large full length fume cupboards favoured by other pharmaceutical companies. The booths are constructed to a zone 1 classification electrically, with the immediate area in front classified as zone 2 (Figure 17).

The dimensions of the booths are 2.5 metres wide, 1.5 metres clear working depth and 2.5 metres high. The booth area is separated into two sections, 014 and 015 (Figure 16). The first section contains 2 booths and the column fume cupboard. The two booths have a removable interconnecting panel to allow the units to operate as one. The second section has a single booth and is suitable for overnight operation if necessary. The booths are designed to give a horizontal laminar flow of air from the front to the rear. Air is drawn from a number of extracts in the rear of the booth. Each of the extracts is fitted with a damper to balance the air flow.

The fronts of the booths are fitted with louvred doors plus vision panels to ensure that the linear air flow is maintained when the doors are closed. The design of these doors caused some problems. The main contractor preferred to test the doors on site rather than build a prototype in a test centre. After considerable smoke testing, the initial design for four sliding doors was rejected and a set of folding doors was developed. With a reduced vision panel in each door and the development of an improved louvre design, smoke testing showed these modified doors were satisfactory (Figure 18).

Air input for this facility is *via* a Burgess tile ceiling arrangement in the zone 2 area immediately in front of the booths. The booths are provided with the standard services supplied to the laboratory fume cupboards plus high pressure steam and compressed air. Compressed air suitable for use with breathing apparatus is also available outside the booths.

The walls of the booths are tiled and fitted with trough sinks on each side approximately 500 mm from the floor slab. These sinks are made from cast epoxy resin for strength and chemical resistance. Floors in the booths are sealed, and coated with an epoxy resin, and are drained *via* a rear gully into a catch tank located towards the rear. The catch tank is arranged so that spillages can either be contained and recovered or made safe and discarded.

The Kilo Scale Synthesis Laboratory also has a large highly ventilated area at its southern end for the use of fan-ventilated steam-heated ovens and steam-heated vacuum ovens. Ventilated hoods are placed above the ovens and above adjacent benching. The benching is specially designed for the handling of materials on the trays used in the ovens.

---

*Electrical equipment in zone 1 or zone 2 electrical areas is designed not to produce a source of ignition for flammable vapours. A zone 1 area must be surrounded by a zone 2 area.

Double ventilated sink and drainer units similar to those used in the standard laboratory are also provided. A separate storeroom is used for the storage of unopened materials and equipment, and a considerable amount of cupboard storage space is available for large equipment both in the laboratory and in the adjacent service corridor.

To ensure safe and clean working practices, showers are located adjacent to the booths. Staff can change into protective garments before entering the booth area, leave their contaminated garments in a dirty zone, and shower before re-entering a clean zone.

6 The Design and Construction of Fume Cupboards for D38

6.1 Development

During the planning stages for D38 the design of the fume cupboards was judged to be the most crucial feature of the New Chemical Research Laboratories at Dagenham. It had been decided that *all* the synthetic chemistry undertaken in this building should be carried out in fume cupboards. It was therefore important that the fume cupboards operated to the highest safety standards and that their design, location, and commissioning conformed to DD80 (later to become BS 7258[1] with the omission of mandatory containment tests). These fume cupboards should reflect their key role in the overall design of D38 by being constructed to the highest engineering standards to ensure their safe and reliable operation. The design of the fume cupboards was facilitated by exploiting the user-group's expertise to produce an acceptable specification. The fume cupboards had also to conform to the RPR standard for fume cupboards developed by the Dagenham Research Centre Building Projects Manager.

The standard liner chosen for the fume cupboards in D38 is constructed of solid epoxy resin. This material is strong, generally shockproof, stain resistant, and minor stains can be removed with fine wire wool. Epoxy resin is ideally suited for moulding into the shapes necessary for the construction of fume cupboard liners. It can be cut to form sinks and grooved to produce sound, strong, and leakproof joints. Liners are made in a one piece, light yellow construction, and are strong enough to allow maintenance and cleaning staff to stand in the fume cupboard when necessary.

Fume cupboards are of the bypass type with a large grill above a closed sash. The extraction plant operates at a single speed giving a fixed volume extract. When the sash is closed, air passes through the grill (Figure 19), but when the sash is opened the grill is blocked (Figure 20).

The user-group decided two types of fume cupboard should be constructed — a standard type and a column type. Both these fume cupboards were designed to our specifications, and mock-up prototypes were built in a test facility by the manufacturer.[7] This initial design was fully tested and developed by the mechanical designer working for the main contractor, the Building Services Engineer and the Senior User Representative from Rhône-Poulenc Rorer.

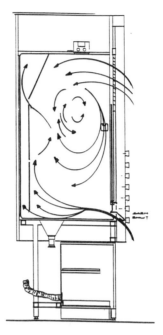

Figure 19 *Fume cupboard air flow with closed sash*

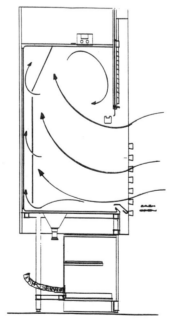

Figure 20 *Fume cupboard air flow with open sash*

After making many modifications during these trials (Figure 21), the prototype fume cupboards were rigorously tested using the sulfur hexafluoride containment test recommended in draft document DD80. These containment tests were carried out by an independent contractor[8] (Figure 22). This test facility also allowed us to examine and develop the air-input flow through the perforated ceiling panels, the extract design, and the effect of the layout of laboratory furniture on fume cupboard performance.

The prototype fume cupboards complied with the air flow characteristics defined in DD80; an average mean face velocity of 0.62 metres/sec was obtained with the sash adjusted to the safe working height for each type of fume cupboard, and containment tests gave results in excess of the minimum acceptable containment index of 3.0-3.5 on the logarithmic scale.

6.2 The Standard Fume Cupboard

6.2.1 Design

The standard fume cupboards are shown in Figure 23. Their dimensions, as specified by the user group, are 2000 mm wide, internally, with a sash safe working height of 500 mm and a maximum sash opening height of 900 mm. The available working depth of 750 mm is more than for many fume cupboards but was also requested by the user-group.

The liner (internal height approximately 1500 mm) has three small sinks cut into its base. Two of these sinks are below the service outlets to the right and left of the fume cupboard, and the third is situated in the centre at the rear. This latter sink is not directly serviced and is primarily designed for use with water condenser outlets. Smoke tests showed the three sinks had little effect on the air flow pattern.

6.2.2 Services

The fume cupboards are all serviced with four pressurized cold water outlets (suitable for water ejection pumps), three compressed air outlets (15 lbs per sq. inch from a central site utility), three nitrogen outlets (from a central bulk liquid utility), two high capacity vacuum outlets (from D38 plant), three steam outlets (from a central boiler house utility) and six individually tripped power sockets (Figure 24). The layout of these services was approved by the user-group. The steam outlets are placed towards the rear of the side panel, and the pressurized water outlets (over the sinks) are aligned with the centre of the sink outlet in order to minimize splashing when the screwed nozzles are replaced by water ejection pumps. The frequently used vacuum outlets are at the front of the fume cupboard on either side.

Two of the three compressed air outlets are located at a higher level, to the left and right of each fume cupboard, for use with compressed air stirrers. It is our normal practice to use air stirrers and not electrically-driven stirrers, especially in the

Figure 21 *Smoke testing of a prototype fume cupboard for D38*

Figure 22 *Containment testing of a prototype fume cupboard for D38*

presence of flammable solvents. Each fume cupboard is fitted with switched socket outlets placed outside the fume cupboard and below the horizontal hinged aerofoil. These electrical outlets are individually tripped to ensure that a hazardous situation is not created due to the failing of another experiment on the same ring main. This will avoid, for example, the failure of an electrical stirrer in a steam heated experiment which might lead to violent bumping of the contents of the vessel if an adjacent piece of electrical equipment 'trips' the circuit.

6.2.3 The Sash

The fume cupboard sash was also developed from the users' specification. Traditionally the sash in fume cupboards at the Dagenham Research Centre incorporated horizontal sliding glass windows. Chemists had become familiar with this design and considered it essential for the new building. On investigation, however, it was realized that working practices had become established in D41 to overcome the poor performance of the extraction facilities. The project team doubted if efficient containment levels could be achieved if we retained this design in D38. We were able to convince the synthetic chemists that with the improved performance of the fume cupboards meeting DD80 standards, sliding horizontal windows were unnecessary.

With the new fume cupboards, the safe working height for the sashes is 500 mm. The sash can be initially raised to 500 mm at which point a solenoid operated locking mechanism operates. This can be switched off by pressing a red button situated in the centre of the fume cupboard, below the horizontal aerofoil, allowing the sash to be raised to its maximum height (900 mm).

6.2.4 Monitoring the Fume Cupboard

The performance of the fume cupboards is continuously monitored by pressure sensors situated in the extract ductwork. Each fume cupboard has an alarm panel situated on the right hand casing (Figure 23). This panel indicates both air flow and sash height status. A green light on the panel indicates normal air flow; any failure and a red light comes on and an alarm hooter sounds, which can only be muted by a key-operated button. Keys to turn the hooter off are kept only by Facilities or Maintenance staff.

Sash height is also continuously monitored. When the sash is between the safe working height (500 mm) and the fully closed position a green light shows on the panel. If the locking button is depressed to release the sash, a red light appears and an audible alarm sounds when the sash is raised above the safe working height. The alarm may be muted by depressing a non-key-operated button and is automatically reset when the sash is lowered below the safe working height. An amber light indicates the reduced extraction speed used overnight, and there is a test button to check that the lights and audible alarms are working. All lights in these alarm panels are fitted with light emitting diodes, rather than normal light bulbs.

Figure 23 *Standard fume cupboard in D38*

Figure 24 *Fume cupboard services positions in D38*

6.2.5 Other Features

Many other design features originated from the prototype fume cupboard in the test centre. For example, the most effective shape for the *vertical* aerofoils was developed to give maximum containment, and considerable effort was also expended on the optimum design for the hinged *horizontal* aerofoil and the front edge protection of the working surface (Figures 25 and 26). The original idea for front edge protection was a rounded protrusion, but smoke tests showed that sharp edges were necessary to ensure that all fumes could be totally contained. This design was tested using a smoke generator, operating at full power in a position 6 inches from the front, and pointing towards the aerofoil. The distance between the hinged horizontal aerofoil and the front edge protection was shown to be most effective at 25 mm. This was to be dramatically illustrated when the finished fume cupboards were containment-tested. The hinged horizontal aerofoil is designed to let all electrical cables and rubber tubing pass under it, and allow the sash to close completely, and thus achieve high containment. Much thought was given to the design of a sash handle which would allow a smooth flow of air and high containment whatever the sash position.

The sash is fitted with toughened glass. Laminated glass was considered, but the shafts of glass produced from this type of glass by an explosion are potentially more hazardous than smaller pieces ejected from a broken toughened glass sash. If a glass panel breaks then the sash is automatically locked. This is essential to ensure that in the event of a glass panel being dislodged from the sash frame, the sudden imbalance in weight between the sash and counterweights does not allow the sash to be violently opened.

The rear baffles are positioned to ensure that the base of the fume cupboard is effectively purged. It is very important that fumes are rapidly removed from the operator's breathing area when a sash is raised. Considerable efforts were made during the development phase to balance the extract vents produced by the fume cupboard baffles, and to create the optimum extraction characteristics over the base, main body and top of the fume cupboard. The baffles are manufactured from 6 mm trespa material, rather than the epoxy resin liner material. The colour of this material is identical to that of the liner, but the cost is considerably lower and these baffles can be easily replaced if damaged. Trespa baffles are also more robust in this suspended state.

The top of the fume cupboard liner is fitted with two twin fluorescent tube lighting units giving a lighting level of 100 lux. These units are situated externally to the fume cupboard liner, and separated by toughened glass panels.

Although it had been agreed that sliding windows in the sashes were not needed, there had to be some provision for additional safety screens within the fume cupboard when a potentially hazardous experiment is being undertaken which requires constant attention with the sash raised.

This problem was resolved by fitting a 'U'-shaped epoxy resin bracket on each side panel of the fume cupboard, just behind the sash and approximately 1000 mm from the base of the liner. When extra protection is needed, a square tubular rod can

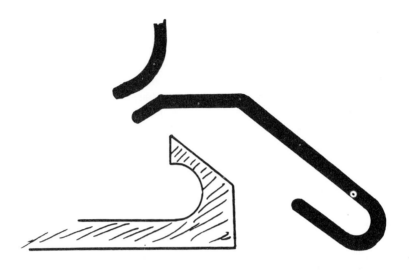

Figure 25 *Fume cupboard base front edge protection design for D38*

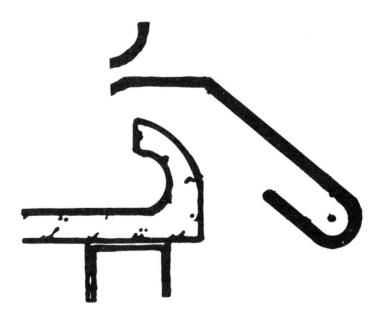

Figure 26 *Original concept for the fume cupboard base front edge protection*

be slotted into the brackets and a flat Makralon safety screen hung from the rod. The safety screen is approximately 500 mm wide, and when supported from the rod allows a space of approximately 30 mm at the base for purging the work surface. It is possible to position the safety screen at any position along the front of the fume cupboard; the front edge protection of the liner base will restrain the safety screen if an explosion occurs in the fume cupboard.

After much consultation and discussion the user-group decided that it would prefer to work with retort stands in the fume cupboards rather than a 'scaffold'. However, fixing points for 'Unistrut' fittings are fitted on the side panels of the fume cupboards and are designed to support a 'Unistrut' frame standing on the base of the liner if it is required.

6.3 The Column Fume Cupboard

6.3.1 Development

When the fume cupboards were being designed there was considerable demand for one with a lower working surface than a conventional fume cupboard. (At the time it was common practice to use relatively long chromatography columns for purifications). It was, therefore, decided to build a fume cupboard with the choice of working surfaces, either 350 mm or 900 mm from the floor. The working surface at 900 mm would be removable for maximum flexibility. This column fume cupboard would be built to the same safety specification as the standard fume cupboard (DD80 recommendations) and have similar features. We soon realized that a 900 mm working surface would be very heavy to remove if cast in one piece and would not be practical. The overall dimensions of the column fume cupboard were again developed by the user-group. The internal liner is 1200 mm wide with a free working depth of 750 mm. From the lower working surface the internal height of the liner is approximately 2000 mm.

6.3.2 Design

The column fume cupboard (Figure 27) has two interlocking sashes and two hinged horizontal aerofoils 350 mm and 900 mm, respectively, above the floor. The fume cupboard is designed to give a safe working height at the upper working surface of 500 mm, and a safe working height from the lower working surface of 1000 mm. A removable sink is an integral part of the upper working surface. To overcome any weight problems, this working surface is built from four interlocking panels of 16 mm thick epoxy resin. The conversion of the fume cupboard from its standard mode (900 mm high working surface) to its column mode (350 mm high working surface) is a simple operation. Firstly, the upper sash is raised and the upper horizontal aerofoil is removed (Figure 28). The lower sash is then raised until it engages with the upper sash. It is then possible to remove the four upper work surface panels (Figure 29).

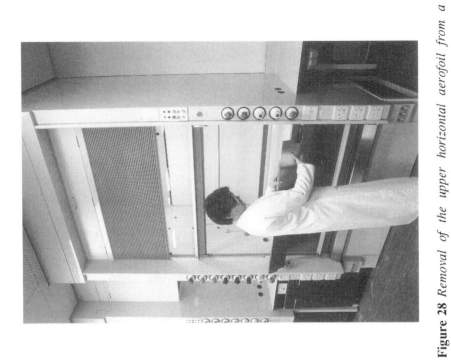

Figure 28 *Removal of the upper horizontal aerofoil from a column fume cupboard*

Figure 27 *A column fume cupboard in the standard mode*

Figure 30 *A column fume cupboard with the high level sink unit in place*

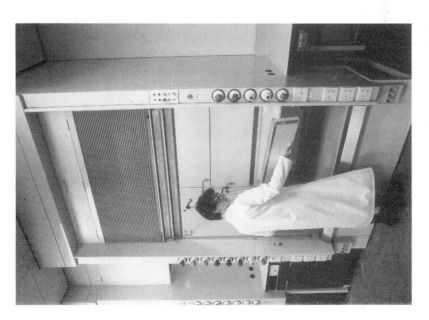

Figure 29 *Removal of the upper work surface panels from a column fume cupboard*

When these work surface panels have been removed, the sink unit remains (Figure 30). The fume cupboard can be used with a raised sink unit or the sink unit can be removed (Figure 31). The interlocking sashes can then be separated and lowered, and the fume cupboard is ready for use in the column mode.

The number and position of the services for the column fume cupboard are similar to those for the standard version. Service controls and alarms are also located in equivalent positions (Figure 27). The sash locking button is on a side panel, and the six individually-tripped switched socket outlets are placed on lower side panels.

During its development, satisfactory containment tests were obtained in both the standard and column modes for this design of fume cupboard. Efficient containment of both chambers in the standard mode is particularly useful when using a vacuum pump in the lower chamber of the fume cupboard, especially when highly toxic materials are being handled.

7 Service Facilities

7.1 Location

Service facilities are located to the south of each laboratory on the ground, first and second floors. The service exit from a laboratory opens out onto a service corridor from which staff have access to a number of support facilities in the service area on each floor, including the cold room on the ground floor (Figure 5, MO41).

7.2 Ventilated Cold Room

Cold rooms used to store chemicals are usually evil-smelling and inhospitable places and we realized that considerable effort was needed to create a clean environment.

Assisted ventilation seemed one answer to the user-group, but effective ventilation of a cold room is difficult. The energy costs are high and condensation problems are likely in summer. After much discussion we decided to limit the ventilation to two or three air changes per hour. The final design of the interior of the ventilated cold room is shown in Figure 32. Open shelves and a closed cupboard unit are provided for the storage of chemicals. The cupboard is for the storage of particularly toxic or smelly materials.

Air is drawn from the cupboard unit into the main exhaust system by a primary fan in a related plant room. This air passes into the cupboard unit from the cold room area *via* a grill in each cupboard door (see Figure 32). Untempered air enters the cold room through a grill in the main door. A drip tray, placed under the ceiling-mounted cooling unit, removes any condensation. A small length of working bench in the cold room is serviced with a switched socket outlet and compressed air supply. The cold room operates at 2 °C, and our aim to provide a clean and safe environment has been successful. All electrical fittings in this area are to zone 1 classification electrically.

Figure 31 *A column fume cupboard in the column mode (sink unit removed)*

Figure 32 *The ventilated cold room in D38*

Figure 33 *A solvent store in D38*

7.3 Service Rooms

A service room, located on each floor in the service area, is equipped with ice-making and glass-washing machines. Most of the glassware used in the laboratories is cleaned in washing machines located in the service rooms. Highly contaminated glassware is initially rinsed with solvent in a laboratory fume cupboard, before it is loaded onto racks located in the laboratory. When the racks are full, they are taken by trolley to a service room and put into the washing machine. The laboratory doors into the service corridor and the service room doors are fitted with delayed action closers to ensure that neither glassware or doors are damaged during this operation. The area above the washing machine is ventilated with a large hood.

Ice-makers in the service rooms are supplied with demineralized water fed from a unit in the third floor plant room. Uncontaminated ice helps to minimize the maintenance of ice-makers. Each service room is fitted with a ventilated sink and drainer unit, and an emergency shower unit. In addition, an argon supply in the service room is used to fill the balloons which provide inert atmospheres for air and moisture sensitive reactions.

7.4 Solvent Store

A solvent store (Figure 33) is located on each floor in the service area. The amount of flammable solvent which may be kept in a laboratory is limited to 50 l by law, and each solvent store is therefore used to store Winchesters of solvent needed for all laboratories on that floor. (Many scientific staff were initially reluctant to adopt this system, and use Winchesters of solvent previously used by others.)

The solvent store is naturally ventilated; high and low level grills allow ventilation to the outside of the building. The service corridor is under positive pressure, and when the solvent store is opened, this excess pressure will cause air in the solvent store to be ventilated outside the building. If the air handling plant which provides pressurized air to the corridor should fail, however, a hazardous situation might occur, and solvent fumes could be drawn from the solvent store into the service corridor by the pressure differential in the adjacent laboratory area.

To prevent this contingency, the air handling plant for the service area has alarms located adjacent to each solvent store. If the plant fails, an alarm rings and staff are warned not to open the solvent store door.

The epoxy resin floor in the solvent store contains a non-slip aggregate; the floor is bonded both to the storage racks (see Figure 33) and to the entrance of the store to stop any spillages leaking into the service corridor.

7.5 Night Running Laboratories

A Night Running laboratory, designed for the unattended running of experiments overnight, is located on each floor in the service area. Each laboratory has two standard fume cupboards similar to those located in the main laboratories. The only minor difference is the provision of a flame detector in the roof of the fume cupboard liner. (Elsewhere in D38 smoke detectors are present in offices, plant rooms, service areas, and laboratories). The fume cupboards in the ground floor night running laboratory are fitted with hydrogen lines from a local cylinder, and this is used for hydrogenation apparatus.

Night Running laboratories are fitted with floor drains and a bonded entrance to prevent any leakage of water. Floor drains are linked to the main sink unit and a supply of water ensures that the U bends do not dry out and release sewer-gases. To avoid hazards related to 'lone working', an alarm is installed, and if a member of staff requires assistance it can be activated, and a bell will ring in the nearby office and writing-up areas.

7.6 Instrument Rooms

Instrument rooms, located on each floor in the service area, are fitted with serviced benches, and house analytical equipment which is not suitable for use in the main laboratories.

7.7 Returns Rooms

Returns rooms are located in the service corridor adjacent to the lift lobby. These rooms contain trolleys which are used for the return of chemicals or empty containers. The trolleys are replaced routinely by Research Technical Services staff, and removed from the building using the specialist electric vehicle described in Section 5.2 (Figure 4).

7.8 Lifts

Building D38 has three lifts: a cable-operated passenger lift (adjacent to the main entrance), a hydraulically-operated goods lift (near the service entrance), and a hydraulically-operated chemicals lift, next to the goods lift.

It is strict company policy that staff must not travel in lifts with chemicals or solvents. The design of the chemicals lift posed a major challenge. The user-group suggested three specifications for the chemicals lift: a zone 1 electrical classification; an over-sized lift shaft should be designed to prevent excessive air movement when the lift moves; the lift shaft should be mechanically ventilated. All these features were incorporated into the design of the chemicals lift.

In addition, the lift car has a tank in the floor to ensure that any spillage is contained. By using a key-operated switch, the lift car will rise to a false floor position approximately 200 mm below the top of the outer door on the ground floor.

In this position it is possible to decontaminate the lift car and the catchment tank from beneath, using breathing apparatus if necessary.

An operating procedure was devised to overcome the problem of placing hazardous materials in a lift and finding that other staff had hijacked the lift before it could be unloaded. When the lift is called, the doors are opened and the chemicals can be loaded into the lift car. Before the doors are closed an internal button in the lift car is pressed, which immobilizes the lift controls on the other floors. The doors are then closed and the lift is sent to the desired floor. It will not respond to another user until the materials are unloaded and the internal button released. If the user does not reset this internal control, the system is designed to reset itself after 10 minutes.

8 Commissioning of D38

8.1 Strategy

The commissioning of a highly-serviced and complex facility such as the New Chemical Research Laboratories, D38, is a long and complex operation. A commissioning engineer should be employed by the contractor and involved from the start of the design process. It is also essential that the user representatives on the project team insist on a high standard of commissioning, and demand that the facility will not be accepted until the commissioning programme is complete. The main contractor will probably not understand the complexity of the operations to be carried out or their related hazards. It is unlikely that a complex facility such as D38 can be commissioned in less than six months. Major problems will be encountered during the commission period and these require both time and resources to rectify.

8.2 Commissioning the Ventilation System

The most important features of D38 are the ventilation system and the related fume cupboards. After the construction of D38, a major programme of ventilation commissioning was undertaken. One of the first problems was the level of noise in the laboratories. The design specification had set a noise level of NR (noise rating) 45 at a position 1 metre from the fume cupboard, but it was apparent that this level of noise was being exceeded when the extraction plant was run for the first time. It was noted, however, that the noise level was not excessive with just one part of the plant operating, but only when the three parts of the extraction plant were in operation — the air-handling input plant, the primary extraction fans and the secondary extraction fans. The extraction system was originally fitted with a single noise attenuator on each fume cupboard extract in the maintenance duct (Figure 8), and a large noise attenuator adjacent to the main secondary fan (Figure 10). It was agreed to fit a second attenuator in the maintenance duct to each primary extract system to try to reduce noise levels.

On completion of this work, the noise level for the ground and first floor had decreased below the specification, but on the second floor was still excessive. A third noise attenuator was fitted to the primary extraction system at plant room level, and this last modification secured the desired noise levels.

The main secondary extract fans were tested, prior to their installation in D38, at the manufacturer's factory in Birmingham. The fans normally operate at approximately 600 revolutions per minute, but for this test the fans were run at 900 rpm with no noise attenuator fitted. This test showed that the 150 kg impeller was unlikely to disintegrate at normal running speeds.

Considerable efforts were made to check that toxic fumes could not leak from the ventilation system. The main secondary fans are fitted with shaft seals to stop leakage of fumes along the drive shaft. The main impeller is fitted with a small impeller on the backplate adjacent to the shaft seal. This small impeller is designed to reduce pressure on the inner side of the shaft seal. Compressed air is blown into the outer side of the shaft seal to ensure that no toxic fumes can be drawn down the drive shaft.

The extraction system was set a severe challenge. Air was blown through a flask containing t-butyl mercaptan in a fume cupboard to produce a significant quantity of a malodorous vapour. This experiment was undertaken in a fume cupboard for every individual extraction system. The fume cupboards chosen were those with their primary exhaust fans nearest to the open louvre of the collection duct in the third floor plant room.

Before this commissioning test began, British Gas, the local police, and the company security department were notified. The test began with electronic and human sniffers located in the maintenance ducts, plant rooms adjacent to the primary fans, in the open plant area on the roof, and at various sites outside the factory. Computer predictions of the exhaust patterns to be expected from the 40 metre chimney (Figure 3) had been supplied by a specialist contractor.

When the tests were carried out, no leakages were detected in the maintenance ducts or from the primary and secondary fan shafts. On some occasions a slight odour was detected in the plant room, and it was discovered that under certain conditions the mercaptan-contaminated air could escape from the open louvres of the collection duct. Following modification to these louvres, to reduce the area of the input and increase the velocity of the air entering the duct, the problem was rectified. Finally, during these tests, no odour of t-butyl mercaptan was detected in the locality of D38.

The safety assessment that was carried out prior to this commissioning test was one of the most complex and exacting COSHH assessments undertaken on the Dagenham site.

8.3 Commissioning the Fume Cupboards

As explained in Section 6.1, prototype fume cupboards were tested using the sulfur hexafluoride (SF_6) containment test recommended in DD80. Containment indices greater than 3.0 on the logarithmic scale were considered to be the minimum acceptable values.

Tests on the prototype standard fume cupboard gave containment indices of approximately 3.5 empty, 3.7 with some small apparatus in the cupboard, and 3.7 with the safety screen suspended from the removable tubular bar. Containment tests on the newly-installed fume cupboards in D38 gave variable results from an unacceptable 1.5 to 4.5. These preliminary tests were undertaken when the building was not completely 'balanced', and a number of factors to explain these results emerged. These included the effect of temperature on containment indices. If the temperature fell below 19 °C at ceiling level, smoke tests showed there was a tendency for air to be dumped on to the floor causing turbulence, before it was drawn into the fume cupboard. At higher temperatures air was drawn smoothly from the 'live' ceiling tiles into the fume cupboard, and higher containment index was achieved. Final containment measurements of 3.7-4.5 were obtained for the fume cupboards in D38.

Another problem identified at this stage were the manufacturing tolerances in the construction of the fume cupboards. The standard and column fume cupboards had been designed with a 25 mm gap between the lower rear edge of the horizontal hinged aerofoil and the upper edge of the liner base edge protection (Figure 25).

Some fume cupboards, however, were supplied and installed in D38 with only a 10 mm gap, and gave low results when tested with SF_6. On altering the gap to 25 mm (by adjusting the support of the horizontal hinged aerofoil), an acceptable containment index was obtained. It is obvious from these results that the fine detail of the fume cupboard design is vitally important and it could also explain the poor results that are obtained by many companies when performing the new containment test in their 'best' fume cupboards.

8.4 The Drainage System

In the New Chemical Research Laboratories, D38, the chemical drainage system is constructed of Schott glass for strength, chemical resistance, and a clear view of any blockages. The number of open drains in D38 has been kept to a minimum, and those that are installed are constantly fed with water to ensure that the U bends remain filled, and no sewer fumes can escape. (This is often a problem in a warm, highly ventilated building, where water in U bends evaporates very quickly). As mentioned previously, there are no floor drains in the standard laboratories, and the U bends in the drains of the Night Running laboratories are supplied with water by the use of the main sink in the laboratory.

The glass chemical drains in the maintenance ducts in D38 pass into a doubly-sealed manhole to ensure that no hazardous fumes are released. Steam condensate water, which is constantly fed to an adjacent drain, also ensures that the drainage system does not leak sewer-fumes. The initial commissioning of the drainage system was rather simplistic. A further series of tests designed by the user representatives was undertaken. These tests included releasing concentrated ammonia solution into the drainage system and mixing it with hot water from the glass washing machines.

No ammonia vapour was detected. Similar tests were also undertaken with acetone and hot water. COSHH risk assessments were undertaken for these tests.

8.5 Miscellaneous

Other facilities which required extensive commissioning were the ventilated cold room, the pressurized water supply, the chemical lift, and the high capacity vacuum system.

Overall our experience has shown the benefit of user representation both in the commissioning programme, and in meeting all specifications.

9 The Building Manual for D38

After D38 had been constructed, and the facilities commissioned to our satisfaction, an induction programme for the staff who were to use the laboratories was begun.

Four weeks before moving into D38, all staff were given a Building Manual. This fifty-page booklet, written by Senior Management and the Senior User Representative, described the main features of the building and how the specialist equipment and plant should be operated. It also included sections on management policy and recommended safe-working practices.

One week before occupation, each section of staff was given a detailed briefing and a tour of the complete facility.

10 Conclusion — Some Minor Problems and Looking to the Future

The New Chemical Research Laboratories D38 facility was opened by the then Education and Science Minister, the Right Honourable John McGregor, PC, MP, in May 1990, and has proved to be an outstanding success with the chemists who have worked in it since its inauguration. We feel that by establishing a user-led project group, we were able to provide a high quality specification. The presence of a very experienced expert user-group facilitated the design, development, and commissioning of the building and ensured no unpleasant surprises in the final facility caused by misinterpretation of the specification.

During the past three years, a few problems have arisen. These have been important, but of a minor nature, and have not interfered with work. For example, software complications in the plant control system caused the alarm system to malfunction, and in the laboratories there were difficulties with the pressure sensor settings. It was possible for a sensor to indicate incorrectly that a fume cupboard extract fan had failed. The result was that the laboratory air input dropped to 50% of its normal volume in order to maintain the negative pressure differential. Normally the sensor would correct itself and the main laboratory air input would be restored to full volume. This often occurred repeatedly and the result was a constant cycling of air by the air input plant. This problem was resolved by adjusting the pressure sensor range in the extract ducting.

Some other minor failures have included the switch-over operation of the high capacity vacuum pumps, the operation of the chemical lift, the setting of fume cupboard balancing dampers, and failure of primary extraction motor bearings. All these faults have been rectified.

When a new building of this complexity is commissioned, it is important to understand that the basic design of the facility is likely to be 2-3 years old. Although considerable efforts are made to incorporate in the final specification all possible amendments to company policy, safety legislation and working practices, it is likely that some aspects of the building design will still be considered unsatisfactory due to the pace of technological change.

For example, when the specification for D38 was being developed it was not company policy to air-condition laboratories. However, following the merger between Rhône-Poulenc Santé and Rorer Pharmaceuticals in the USA to form Rhône-Poulenc Rorer this policy has softened. Should we now fit cooling systems in the laboratory areas? Is it practical?

When the specification for the laboratories was being considered, the standard method for the purification of crude mixtures of products used long chromatography columns, and hence a column fume cupboard was needed. The change to 'flash' chromatography (which uses much shorter columns) means there is less demand for a purpose-built fume cupboard for chromatography. These column fume cupboards are also used for large-scale synthetic chemistry, and the handling and distillation of highly toxic materials.

The Rhône-Poulenc Rorer project team set out to design a facility that will allow the Dagenham Research Centre to meet its scientific objectives this decade and into the 21st century. This task has been achieved, and the scientists at Dagenham have both the equipment and the laboratory environment with which to exploit contemporary synthetic chemistry and pursue new drugs.

11 References

1. British Standard 7258, Laboratory Fume Cupboards, Parts 1, 2, 3, British Standards Institution, London, 1990.

2. Haremead Ltd., The Old Mill, Mill Lane, Godalming, Surrey, UK.

3. Professor John B. Taylor, Senior Vice President, Central Research.

4. Mike Spencer, Dagenham Research Centre Building Projects Manager.

5. Susan Hills, Malcolm Palfreyman, Chris Smith, Trevor Harrison, Alan Thatcher, Malcolm Toft.

6. J.A. Elliot Ltd., 133 Stansted Road, Bishops Stortford, Essex, UK.

7. PF+F (Radcliffe) Ltd., Booth Road, Little Lever, Bolton, UK.

8. Modular Surveys Ltd., 25 Milton Drive, Poyton, Stockport, Cheshire, UK.

Subject Index